Rogue Wave:
The U.S. Coast Guard
on and after 9/11

For sale by the Superintendent of Documents, U.S. Government Printing Office
Internet: bookstore.gpo.gov   Phone: toll free (866) 512-1800;   DC area (202) 512-1800
Fax: (202) 512-2250 Mail: Stop SSOP, Washington, DC 20402-0001

ISBN 0-16-069824-3

ROGUE WAVE:
The U.S. Coast Guard
on and after 9/11

by Chief Petty Officer P.J. Capelotti, Ph.D., USCGR

Foreword by Admiral Thomas H. Collins, Commadant, USCG

U.S. Coast Guard Historians Office
Dr. Robert Browning, U.S. Coast Guard Historian
U.S. Coast Guard Headquarters (G-IPA-4)
Washington, DC 20593
www.uscg.mil

Contents

FOREWORD

On 11 September 2001, more than three thousand innocent people died when terrorists hijacked four commercial airliners for use as weapons against the people of the United States. The strikes against the World Trade Center and the Pentagon were calculated, murderous acts unmatched in our nation's history. Men and the women of the Coast Guard, including reservists and auxiliarists, reacted swiftly and courageously. As a Service, we met the immediate challenges put before us with purpose and shifted focus to address this new terrorist threat to our homeland.

The 9/11 disaster revealed to the Coast Guard, and to the nation as a whole, the new scope of operations that are necessary to keep us safe. Many lessons were learned that day and in the post-event analysis. These lessons helped the Service to refocus on one of its fundamental responsibilities, the security of America's maritime frontier. This mission is essential, not only for the protection of our citizens, but also to ensure the smooth flow of our nation's vital commerce.

This responsibility is multi-mission in scope and is best carried out through the skills and esprit de corps of a military organization. When disaster struck on 9/11, America witnessed first hand the Coast Guard's agility and flexibility. Resources were promptly shifted from a multitude of other missions to rapidly increase the security of our ports. In the ensuing days, weeks and months the nation came to put its faith in the men and women of the U.S. Coast Guard, on watch along our coasts, lakes and rivers and around the globe.

Recording our history is something that we haven't always done very well in the past. The lack of resources and the general feeling that our deeds and performance spoke for themselves caused us to ignore the genuine need to capture our history.

This shortcoming was recognized even before 9/11, with plans and proposals aimed at growing the Coast Guard's history program already on the table. VADM Howard B. Thorsen, USCG (Ret.), a stalwart supporter of the need to capture and document Coast Guard history, has been instrumental in sensitizing Coast Guard leadership to this need.

It was with this heightened sensitivity that the Office of Governmental and Public Affairs mobilized reserve Chief Petty Officer Capelotti and challenged him to record the history of what the Coast Guard did on 9/11 and throughout the days and weeks that immediately followed. Chief Capelotti has done a superb job capturing the Coast Guard's response. His work will prove to be valuable reading for current and future leaders in the United States Coast Guard for years to come. And it is a first step of many to strengthening our Service's commitment to recording the stories and the lessons of who we are and what we do.

Admiral Thomas H. Collins
Commandant, U.S. Coast Guard

Preface (Pentagon): Rogue Wave

"Senior officers, often unaware of the impression they made, set examples their men would never forget."

-Cornelius Ryan
A Bridge Too Far

Many years later, on the morning of September 11, 2001, Jeffrey Hathaway remembered well the best job he ever had in the U.S. Coast Guard. As a young Lieutenant Commander, he had assumed his first command at sea, a 180' buoy tender named *Citrus*. The old vessel had been painted white and converted to a Medium Endurance Cutter based out of Coos Bay, Oregon. It was November, just before Thanksgiving. The crew of the *Citrus* had been scheduled to be home over the holiday, when a storm roared up over the Oregon-Washington coast. Many fishing boats were caught offshore when they couldn't get back in over the bar. Hathaway decided to keep the *Citrus* underway, to stand ready in the event that any of the fishing fleet trapped offshore in the storm called for assistance. He told his crew they would celebrate the holiday at sea, albeit in the absence of a turkey dinner.

Riding out the storm, not going anyplace in particular, just station-keeping, *Citrus* was sliding at an angle down the back side of a swell when Hathaway heard the voices of his crew on the bridge screaming "oh my God." He stood and looked out one of the two portholes in his cabin that gave a view out onto the buoy tender deck. A giant wall of water was heading straight for his cutter. It was a rogue wave, twice the height of the swells around it, coming on at a 45 degree angle from the others, as tall as the masthead of the *Citrus* itself. The rogue swell broke all the way over the mast, carrying away the radar antennas. Hathaway was knocked down, but got up and made his way out of his cabin and up to the bridge. He expected to see daylight above him, believing that the whole superstructure had been carried away.

Instead, the bridge was intact. The bridge crew were picking themselves up off the deck. Hathaway looked out in front of *Citrus* to see the wake of a ship, along with a whole field of debris in the water. His first thought was that another vessel had sunk. Then he realized that the rogue wave had taken his 1,800-ton Coast Guard cutter and moved it backwards several hundred yards, and *Citrus* was now steaming slowly back through its own debris field. Aluminum ladders had been torn off, all the vessel's small boats had been ripped from the decks and were now tossing in the sea, and life jacket lockers had exploded. Kapok jackets stenciled with "Coast Guard Cutter Citrus" bobbed around the vessel.

Then the phone started ringing. First came a report of flooding, from hatches that had blown open, exposing the lower decks to the sea. Then, worse, a report of fire, as electronics started shorting out. For Hathaway, it was a mariner's worst nightmare: flooding and fire at the same time. At that moment, standing on the bridge, he suddenly realized that everyone on the bridge was looking at him, and that his crew had only one question, a question only he could answer. They all wanted to know: what was the Captain going to do?

At 0943 on 9/11, terrorists piloted American Airlines Flight 77 into the Pentagon, destroying the U.S. Navy Command Center under the command of now Rear Admiral Jeffery Hathaway. Forty-two people were killed instantly, including twenty-seven who worked directly for the admiral. In the days and weeks that followed, he found himself at the head of dozens of funeral caissons as they made their way through Arlington National Cemetery, in full view of the devastated sector between Wedge 1 and 2 of the Pentagon where the people he was burying had died. At the same time, he had to see to the survivors, oversee an interim command center and the rebuilding of a new command center: all while coordinating the anti-terrorism and force protection policies of the U.S. Navy, around the globe. His face was like a chart, one that marked the channel leading from pain to resolution.

The same thoughts that he had experienced on the *Citrus* came back to Hathaway after the attacks of 9/11. Survivors were scared, some looked to escape, no one knew if another attack was imminent. And everyone was looking at him, looking for someone to tell them that it was going to be okay, to say that 'here's what we're going to do,' and then to lead the way. It was very likely the most difficult moment of Hathaway's career as a Coast Guard officer, but he had some experience to draw upon. He knew how to rally a frightened crew, how to put a command back together after it had been knocked down, knew where to go and how to get there. Most of all, he knew a rogue wave when he saw one.

Preface (New York): *41497*

Carlos Perez was at loose ends. He had been enrolled in a community college in Brooklyn for a couple of years, but found himself attending too many parties. He worked at a local gym but wasn't satisfied with that. Watching television one afternoon, an advertisement for the U.S. Coast Guard caught his attention. That looked like something he might want to do. The next thing he knew, he was standing in formation at the Coast Guard Training Center at Cape May, New Jersey.

Four years later, after being stationed on the cutter *Nunivak* in Puerto Rico, Boatswain's Mate Third Class (BM3) Perez found himself on Staten Island, at Station New York. Duty sections were broken down into three areas: watches, law enforcement, and search and rescue (SAR). Perez had qualified as a coxswain about three months earlier, and on that early September morning was the coxswain in charge of Coast Guard utility boat *41497*, the "ready boat," poised to get underway at a moment's notice in the event of an emergency. The sky was beautifully clear and New York Harbor smooth as glass.

On that Tuesday morning, Perez was on search and rescue. Perez was in the chart room at the station, correcting his charts, when he heard the SAR alarm sounding. For a brief moment he shrugged it off; they had been having problems with the SAR alarm. He went outside, and as he saw smoke coming from one of the World Trade Center towers, the officer of the day ordered *41497* to head to the scene.

A Coast Guard utility boat can transit from Station New York across New York Harbor to the Battery, the southern tip of Manhattan Island, in about fifteen minutes. Like most everyone else, Perez and his two crewmembers on *41497* assumed the tower fire to be an accident. Perez had watched commercial aircraft heading for LaGuardia by circling near the World Trade Center for much of his life, so it was not unthinkable that one had strayed horribly off course. As a kid, he had often thought about the catastrophe that would follow if one of those towers fell. Many days later, when he had a moment to think about it, he realized that all of the bad things he had pictured in his imagination as a child really did happen.

They had just reached the tip of Governor's Island when Perez' lookout, Petty Officer Landon, told Perez to look out his window. They sighted a plane nearby, and it looked like it was flying too low. Perez craned his neck to get a view out of the port side window, just as a large aircraft swooped directly over his boat. Perez looked up to see the underbelly of a commercial jetliner. What he saw was United Airlines Flight 175, flying from Boston to Los Angeles, with a crew of nine along with fifty-six passengers. Five of those passengers were suicidal religious fanatics intent on mass murder. The aircraft screamed directly over Coast Guard *41497* and smashed into the south tower of the World Trade Center.

Perez stood at the helm, stunned. He remembered thinking that it was like something from a Spielberg movie. He looked at his two crewmembers and they looked back at him. "We had a long moment of silence, a long moment of silence." How did one react to such a sight? Especially since Carlos Perez was no mere onlooker. It was his very first search and rescue case as petty officer in charge of his own boat, and the distress call he was answering just happened to be 9/11.

He radioed back to Station New York to report that a plane had hit the towers. If the nation were under attack, his superiors would want information that only *41497* could provide. But all communications had suddenly gone crazy. It became impossible to reach his command. Perez put aside the mike. He had bigger problems. He turned his attention and his boat to the people of New York; where they were going and how they were going to get there.

It was in many ways a classic Coast Guard situation, one that will confront almost every member of the Service at some point early in their careers. A young sailor, with responsibility far beyond his rank or rate, in charge of a boat and a crew, cut off from his command, and, for the moment, the only U.S. Coast Guard presence in the harbor, the only law on the water.

Perez snapped free from his shock. Faster than those around him, Perez recognized that his country was under attack and understood the consequences. A Coast Guardsman is both Guardian and Lifesaver, and Perez now looked to both traditions in his response.

First he directed his crewmembers to keep a sharp lookout, 360 degrees, on the water and in the air. They had been attacked through the air. An attack by a ship might be next. Then his instincts as a lifesaver took over. As smoke and debris started to fill the air, Perez drove *41497* directly towards it, towards Battery Park. He began a series of sweeps of the waterfront, in the lookout for people in the water, for debris that might be used as evidence in the disaster. He continued to keep an eye on the towers, and through his binoculars he watched as people jumped to their deaths from the upper floors.

An hour later, driving the boat past the Staten Island ferry terminals, the crew of *41497* heard a low rumble from the area of the Trade Center. As the rumble intensified he watched the South Tower splinter apart and come down. "A dust cloud like some kind of monsoon started rolling towards us, and it became so dark that you couldn't see anything," so Perez engaged his engines and stood off from the Battery, trying to assess the situation. Paper and other debris flew everywhere. The whole area from the towers to the harbor south of Governors Island was blanketed in a dust cloud. When the North Tower fell, another wave of debris, like a fog bank, rolled toward the harbor.

Perez headed *41497* back into the thick smoke, searching for people who might have tried to escape the maelstrom by jumping into the harbor. To Perez it seemed as if the whole population of Lower Manhattan had suddenly gathered inside a dark dust cloud at the Battery, desperate to flee the island. There was very little talk amongst the crew. Their thoughts were focused on staying in position in the event they were needed to save lives in the water. Even though the station had a weapons locker, on that morning *41497* carried no firearms. Perez had been trained that when the SAR alarm went off, he was to grab his float coat, get his crew on board, and rush to the scene and offer any assistance required. But now he was faced with a situation that confounded Coast Guard search and rescue doctrine, a deliberate attack upon an undefended civilian target. Beyond response and rescue, he now had a third mission to consider: public reassurance.

By late morning, every asset the Coast Guard could throw at the disaster was either in the harbor or en route at flank speed. Scattered communications were coming back on line and Perez and other small boats were directed to sweep the island, rendering assistance wherever necessary. It became impossible to see anything. Ferries and tugboats crowded the harbor and amid the smoke and dust Perez used his radar to navigate his way around them. Another small boat from Station New York, a fast rigid hull inflatable boat (RHI), came alongside *41497*. Perez briefly turned *41497* over to one of his crewmembers while he and the petty officer in charge of the RHI, Boatswain's Mate First Class Kenneth Walberg, returned to Station New York on the RHI. There they grabbed a bagful of dust masks and goggles and returned to the disaster. Through the dust, Perez could see the fear in the eyes of the crowds gathered to make their ways back to their homes and families.

But as they closed on the Battery, Perez could already see the result of his third mission that morning, could sense the effects that the appearance of a U.S. Coast Guard vessel had on stricken Americans. He watched as they responded differently to his and other Coast Guard small boats and cutters than they did to commercial or private vessels. He saw them gain a sense of reassurance and ease from the very presence of a Coast Guard boat as they sought to escape from the city.

In the two months that followed 9/11, Carlos Perez and the other coxswains of the U.S. Coast Guard drove themselves and their small boats to the limits of endurance. They put the equivalent of twenty-two *years* of use on their boats, patrolling and securing the harbors of the homeland. The Coast Guard itself would be transformed from a low-profile, low-budget fifth armed force, into the principle defender of the ports and waterways, the bays and banks, of the American homeland. It would find itself with a billion more dollars, and on its way into a whole new federal agency, its historic anonymity gone forever.

But for Boatswain's Mate Perez, he would always remember a beautiful morning that didn't stay beautiful, and a plane coming hard over his shoulder, and his first search and rescue case on board *41497*.

Chapter One

9/11 at Activities New York: Richard Bennis and the Performance of an Activities Command

"In Dick Bennis, we had the right guy at the right time in the right place."

-Admiral James M. Loy

Prior to September 11, 2001, Richard E. Bennis held the distinction of having served as a Captain of the Port (COTP) in three of the largest container ship ports in the U.S.: Charleston, South Carolina; Norfolk, Virginia; and New York. He had also held the position as at Headquarters which put him at the head of policies for the Coast Guard's response to explosions, fires, natural disasters, and oil spills, and was co-chair of the national response team, an interagency coordinating body of all federal response organizations. In the field, he had been part of the security team for the 1996 summer Olympics and the *Exxon Valdez* spill. He had been in Norfolk for less than a year when the COTP job in New York came open, and with the impending arrival of OpSail 2000, the largest peacetime gathering of sailing ships since the rededication of the Statue of Liberty in 1986, Bennis saw an opportunity for some interesting work.

A native Rhode Islander, Bennis arrived in New York to an expected tempo of incidents, and these helped sharpen the planning underway for the International Naval Review and OpSail 2000. OpSail involved not only a large Coast Guard security evolution to protect the gathering of visiting sailing vessels, but numerous partnerships with state and local law enforcement agencies and commercial shipping concerns. In the wake of the success that followed OpSail 2000, Bennis was selected for flag rank, and extended for another year at Activities New York.

At the time, Bennis looked forward to a year of writing his lessons learned from OpSail, while anticipating what new assignment the Coast Guard might ask him to perform. It was then that he learned he had been diagnosed with incurable melanoma, and the cancer eventually invaded both his lungs and his brain. Refusing to listen to the prognosis that he had six months to live, the Admiral underwent eight hours of brain surgery, had a metal plate emplaced in his skull, and a day and a half later was back at the office.

For Bennis, OpSail 2000 became a kind of trial run for an even bigger security nightmare that would follow one year later, one in which the Admiral emerged in many quarters as a skilled crisis manager and a beloved friend of the enlisted ranks.

> We mentally dusted off all the lessons learned from OpSail: how to support a large influx of Coast Guard personnel, in particular how to support to PSUs (Port Security Units), berthing, transportation, fuel. With that many small boats, between the PSUs and the Auxiliarists, we had major issues with regard to messing, fueling, berthing, logistics, the whole range. We knew that the last time they had been there the PSUs didn't have enough sinks and mirrors and electrical outlets for shaving. All of those essentials: showers, telephones, head facilities, these were things that seamlessly on 9/11 we transitioned into from the two years we had spent preparing for OpSail 2000. 9/11 gave us no time for preparation, so [using that earlier experience] we just made it happen.

Bennis had been one of the officers on the Activities Evaluation Team at Headquarters when the concept of a unified Group-MSO command was studied. Different models had been tried in various locations around the country, from Corpus Christi to San Diego to Baltimore to New York. It was in fact Bennis' interest in how the model was working in New York that caused him to seek that position, as commander of the largest field command in the Coast Guard. He liked the Activities New York model of one-stop shopping for both the community and for the Coast Guard, itself, for the many services the Coast Guard was called upon to provide in a major port environment.

Because of the Activities consolidation, the entire port community knew where to find the Coast Guard in New York. Unlike other ports, where someone who needed the Coast Guard might go to two or three different locations before finding the marine safety or operational or public affairs person they were looking for, in New York one only need go to Activities New York at Fort Wadsworth on Staten Island. When people would ask how he could coordinate all the Service's disparate functions in one place, Bennis responded that the opposite was true. Activities New York simplified his job as Captain of the Port of New York almost to the point where it was easy, the very opposite of other ports where Bennis had worked earlier in his career. It was hard to explain to people outside the Coast Guard, but without the Activities concept, the COTP, without any operational assets, was often in the position of having to liaison with his own Service in order to get a ride across the harbor.

> I could get a call in the middle of the night from our response team or our investigators, saying, for example, this vessel has just hit a bridge. There's a fire, there might be an oil spill; we think there are people in the water. We've got to send somebody out there to do investigation, to gather evidence, and we've got to send people out to respond to the possible oil spill. The Group has already sent their rescue boat out; they don't want to send their stand-by ready boat out, so we've got to get a ride from somebody else—maybe one of the [civilian commercial] harbor pilots can take us out there, and we'll keep you posted. They'd call back in half an hour saying they were having trouble getting across the harbor, and can you call the Group Commander and get him to free up his ready boat, because the crew is uncomfortable releasing it. That was the norm.
>
> Now, in New York, I get one phone call: an event has just happened, a boat is underway with SAR folks and members of the response team, and we've got a second boat standing by for the investigators and they'll be going out shortly. That was an absolute joy for me.

Enlisted members, or junior officers whose first experience was at Activities New York, would often return from a stint at another unit where such Marine Safety and Operations—so-called "M" and "O"—questions were still fought over, and marvel at the wasted effort. In New York, a call came in, and Activities responded. That was it. "We don't have an "M," we don't have an "O,"" said Bennis. "We just have the Coast Guard, which has missions, and which goes out and responds."

Prior to 9/11, the operational assets attached to Activities New York gave it a sizeable fleet with which to create a rapid on the water presence that morning. These included two 140' Bay Class icebreaking tugs, *Penobscot Bay* (WTGB 107) and *Sturgeon Bay* (WTGB 109); two 110' Island Cutter Class patrol boats, *Adak* (WPB 1333), and *Bainbridge Island* (WPB 1343), along with three 65' small harbor tugs, *Hawser* (WYTL 65610), *Line* (WYTL 65611), and *Wire* (WYTL 65612); a 47' motor lifeboat, and a host of 41' utility boats (UTB) and rigid hull inflatables (RHI) from four separate small boat stations. These were complimented by the 175' Keeper Class coastal buoy tender *Katherine Walker* (WLM 552), which, while controlled by First Coast Guard District in Boston, was considered part of the Activities family.

*   *   *

On September 10th, 2001, Admiral Bennis had staples removed from the back of his head from the cancer surgery he endured earlier in the summer. He and his wife thought that maybe the time had come to slow their lives down, so they decided to head south to look for a retirement home. He asked his doctor if it was okay to travel the next day, and given that blessing the Admiral was on his way early in the morning of September 11th.

In Bennis' absence, his deputy, Captain Patrick Harris, convened the usual morning staff brief at 0830 in its usual spot, the crisis action center. A career aviator, with thirty-four Air Medals from his Vietnam service as an Army helicopter pilot, Harris now balanced command responsibilities with all of the combined "M" and "O" divisions he supervised in the port, everything from a Vessel Traffic Service (VTS) to small boat stations to a pair of icebreaking tugs. The "activities concept" had compressed almost half a dozen commands that once existed on Governor's Island and around New York Harbor into the single command at Fort Wadsworth. Harris usually awoke with a lot on his plate, but remarked that Activities provided a lot of tools to solve them.

The morning brief had just concluded, and Harris was preparing for an awards board, when the watchstander came in from the operations center and said that one of the towers was on fire. Harris made his way the few steps to the ops center, where he saw that not only was Cable News Network (CNN) showing the fire at the tower, but a dozen of Activities' sixteen strategically-placed VTS cameras were trained on it as well. Harris stood at the SAR desk, watching the television broadcasts, as reports began to come into the ops center from tugs in the harbor. As an aviator, Harris was uneasy with initial reports that a light plane had accidentally hit the Tower. Even though the incident was so far a land-based one, he decided to position his forces in case they were needed.

Station New York launched its ready boat, a 41' utility boat under the charge of Boatswain's Mate Third Class Carlos Perez. Admiral Bennis was on his way south, driving through northern Virginia on I-95 near Quantico, when Harris called him. "You've left town again, Admiral," Harris said. "And something always happens when you leave town." The last time Bennis was away, someone trying to pull a stunt had gotten their parachute stuck on the Statue of Liberty's torch. It was the kind of thing that happened a lot in New York. Harris asked the Admiral to get to a television, since a plane had just hit one of the towers of the World Trade Center. Bennis then spoke with Rear Admiral George Naccara in Boston, and between them they decided that Bennis needed to turn around and get back to New York.

At 0903, the second plane hit. Perez and the few other Coast Guard small boats in the harbor saw it happen in front of them. Captain Harris was watching it on a television in ops. "At that point," remembered Harris, "we knew what we had." Seven minutes later, at 0910, he closed the port. "All our vessels immediately moved in toward the scene. Any vessels not underway were put underway within minutes. Very shortly, we had launched everything. We had one vessel in "Charlie" status [at the dock for routine maintenance and crew leave], and even it was moving within two hours. We had volunteers start showing up within minutes of the second plane hitting. We turned the entire unit operational in minutes, and supported all the operational missions with our marine safety folks. They turned operational. We turned into a true response organization for about six weeks."

Lieutenant Sean MacKenzie, a 1992 Coast Guard Academy graduate, had just started to settle into his first command assignment as skipper of the 110' Island Class cutter *Adak* on 9/11. *Adak* was at her berth at Station Sandy Hook, getting her steering system fixed. The steering mechanism was completely disassembled when the first aircraft hit the Trade Center. From his perch on the cutter's open bridge, MacKenzie could see the smoke coming from the towers. His summer of fisheries patrols off Cape Cod, accompanied by beautiful weather and flat calm seas, suddenly seemed like a far-off dream, as Activities New York called the Station and asked *Adak* to make preparations to get underway. Electrician's Mate First Class Juan Vasquez jury-rigged the *Adak's* steering system so that MacKenzie could steer the cutter by hand.

Adak's lines were taken in and the cutter backed out from her berth. Spouses of the crew, alerted to the sudden departure, waved from the pier. *Adak* made her way into the Sandy Hook Channel and toward the Verrazano Bridge, twenty-five minutes away at full speed. From the bridge, *Adak* ran across New York Harbor and at noon took up station off the Battery at the southern tip of Manhattan. Quartermaster First Class Matthew James placed a piece of Plexiglas over *Adak's* chart of the harbor, and he and Lieutenant MacKenzie began to plot the positions of all of the Coast Guard cutters and small boats converging on Lower Manhattan. From that moment, until the arrival after midnight of the *Tahoma* from New Bedford, Massachusetts, *Adak* acted as a command and control center and On Scene Commander for all Coast Guard underway units in New York, to keep everyone out of the harbor save for the vessels directly providing rescue and assistance.

In Virginia, Admiral Bennis had retrieved a small battery-operated TV from the trunk of his car, which he handed to his wife as he turned the car around to head back to New York. "Oh, here's an instant replay," said his wife, adjusting the antenna to get better reception. It was the second plane. Bennis put his hands-free earphone on, and sped north on I-95. Just as he crossed the Woodrow Wilson Bridge across the Potomac, he looked upriver to see smoke coming from the area of the Pentagon. From the Wilson Bridge, Bennis accelerated to 95 mph, often using the breakdown lane, driving straight from Washington, D.C., to Station Sandy Hook, New Jersey, where Captain Harris had arranged for a boat to be ready to meet the admiral.

Bennis was met by a Coast Guard Reserve coxswain who had been driving across the George Washington Bridge when the first plane impacted. He had immediately changed his destination and, without being called, reported to his unit at Sandy Hook. As the boat made its way from the spit of land at Sandy Hook, past the Verrazano Bridge, with its view of Manhattan which had always been so awe-inspiring and was now full of smoke and fire, Bennis saw all of the crew had tears in their eyes.

Captain Harris wanted a 270' medium endurance cutter on scene as soon as possible for command and control. He had issued the order that closed the port, and needed a Coast Guard presence that would symbolize to all potential trespassers that the harbor was in fact closed completely. Activities requested, as well, forty-two Tactical Law Enforcement (TACLET) personnel. TACLETs are an outgrowth of the Coast Guard's Law Enforcement Detachments, better known as LEDETs, formed in 1982 to deploy on U.S. Navy ships in support of counter drug law enforcement. The once-scattered LEDET's have been consolidated into three TACLETs located in Yorktown, Miami, and San Diego. With all inbound commercial vessels stopped at the Ambrose Light, TACLET teams would begin boarding every vessel attempting to enter the harbor.

If a terrorist vessel was already in the harbor planning to ram a bridge or maneuver into position for some other kind of attack, the only way to defend against it was to close the port and stop all traffic. Only a large cutter could do that. Harris separated the forty square mile harbor into security sectors, and when the 270' Bear Class cutter *Tahoma* (WMEC 908) arrived, it took over the control of the small boats in the harbor. TACLET personnel were put on board small boats, as well, to stop and search any suspicious vessels within the harbor. Harris also retrieved the plans that Activities had used for OpSail 2000.

> Within two hours of the attacks we broke into our OpSail/INR [International Naval Review] 2000 plans, and looked for everything we could find that might be of use. That became, first, points for a mass evacuation of Manhattan, and, second, sectors we had used for controlling traffic. We started to do that but it was too widespread. Those twelve sectors were supposed to have seven or eight vessels in each sector. We didn't operate that way on 9/11. We didn't want so much to control vessels that wanted to come and sit, but vessels that wanted to go through and look. So we had to set it up differently.

Besides the security sectors, the Activities staff used the last part of the OpSail 2000 evacuation plan for lower Manhattan. "We used the evacuation sites of where things went to, not where they went from," remembered Captain Harris. "It was a very concentrated evacuation, with everything coming from Lower Manhattan. We hadn't planned it that way. That day we had buildings falling down, debris coming down, and dust and soot in the air. We had people running for the first piece of shoreline or the first thing that they knew of."

By mid-morning, the Coast Guard had nearly forty boats operating in the harbor, along with a growing fleet of private and commercial vessels that were trying to help. The tug *Hawser* was in Charlie status on the morning of 9/11, at its berth at the Military Ocean Terminal (MOTBY) in Bayonne, New Jersey, being refueled after having returned the previous evening from an upriver patrol to Waterford, New York. *Hawser's* skipper, Chief Boatswain's Mate James A. Todd, was having breakfast. Todd had been in the Coast Guard for fourteen years, ever since the day he went to lunch from a boring job at a clothing store in Charlotte, North Carolina, noticed a Coast Guard recruiting station across the street, walked in, and never looked back. Now a Chief Boatswain's Mate and Officer in Charge of the 65' harbor tug, Todd's responsibilities reflected the multi-mission nature of all Coast Guard platforms, and included marine safety, law enforcement, search and rescue, and some aids to navigation work. *Hawser's* patrols had taken it from Sandy Hook, New Jersey, to—with the top stripped of antennae and the Hudson River down at least nineteen inches to allow *Hawser* to clear the bridges—as far north as Lake Champlain, the first Coast Guard tug to operate on the lake.

When the first plane hit the Trade Center, Chief Todd immediately headed back to the tug and recalled crew members who had just left for liberty. Within minutes, Todd ordered the tug and its compliment of seven crew members to head for lower Manhattan.

> We headed right for the buildings. We knew that there would have to be a security zone set, we knew there would be a need for medical facilities to be set up and people to be taken off the island. We knew we had a job even before being contacted. There was a mass of people on the shoreline... We took some people off, passed out gas masks, and unloaded our water, so people had potable water. We directed people to the medical centers that were set up, and tried to maintain some sort of order with the [commercial] tugs coming in to take people off...

Communications were being maintained along diminished pathways, both through low-site VHF radio and through sporadic cell phone communications in the VHF gaps. Responding to reports that a frantic evacuation was underway, Captain Harris sent Activities Marine Inspectors "with good strong command voices" away from Fort Wadsworth *en masse*. When the south tower collapsed, Vessel Traffic Service at Activities had put out a call for any vessels that could respond to come to the docks and assist the situation, and coordination of that diverse "fleet" would be a monumental task.

The inspectors went to a pilot boat and to several ferries, and moved out toward Manhattan to provide waterside security and direct the maritime evacuation. They calmed nerves on board the ferries, and in many places got off the ferries in Manhattan to stand at the head of piers and direct people to the appropriate waterside exit, be it to New Jersey or Staten Island or Brooklyn. And, being inspectors, they made certain that before a ferry left its dock that it was not overloaded.

One of those inspectors was Lieutenant Michael Day, the Chief of Activities New York waterways oversight branch. Before coming to Activities New York, Day had actually worked for the Port Authority of New York and New Jersey in 1 World Trade Center, the north tower, as a member of the Coast Guard's industry training program. He had also served on board the Bear Class cutter *Seneca* (WMEC 906), during Operation *Able Manner/Able Vigil*, the interdiction of Haitian and Cuban migrants in the Strait of Florida and the Windward Passage. He would later look back on those earlier mass migrant evacuations in the Caribbean in part as preparation for what he would face in Manhattan on 9/11.

On the morning of 9/11, Day had a previously scheduled meeting in Manhattan, and had planned to carpool with a Sandy Hook pilot named Andrew McGovern. McGovern now offered a 200' pilot boat, moored at a dock next to Station New York, for the use of the Coast Guard in the growing emergency.

Lieutenant Day, along with Chief Petty Officer Fred Wilson and a few other petty officers, boarded the Sandy Hook Pilot Boat *New York*. "It was chaos," remembered Day. "Every channel you clicked to, people were screaming, 'Help! I need people over here! I've got someone hurt here!' Everyone was talking over everyone else. Cell phones stopped working." In the growing absence of communications, rumors abounded—martial law was about to be declared; suicide bombers were loose in the city—and they had to be dealt with and eventually shut out.

Proceeding first to the Battery, Day asked at the Coast Guard building there for a Coast Guard ensign, along with some ball caps and float coats. The ensign he raised on the *New York*. The caps and coats were used to reinforce their Coast Guard presence on the pilot boat. Then Day began talking on the radio, broadcasting from the pilot boat. "We got on the radio and said, 'United States Coast

Guard aboard the pilot boat *New York.* All mariners, we appreciate your assistance.'" He proceeded to ask the converging vessels to head for marshalling areas, and avoid clumping together at over-crowded dock facilities.

Day and Wilson began to direct and coordinate the evacuation of lower Manhattan by water just as the North Tower of the Trade Center collapsed at 1030. They pressed into service a civilian marine inspector with intimate knowledge of the capabilities of the harbor ferries, giving him Coast Guard coveralls, a ball cap, and a float coat. Using their expertise to separate the chaos of tugs and ferries, the Lieutenant and the Chief directed vessels to critical routes based on their size and maneuvering characteristics.

Proceeding to several docks around the tip of Lower Manhattan, Day dropped off petty officers at each, to work with police and help settle crowds and keep the waterside evacuation under control. Day remembered people cheering as their over-loaded ferries left the docks, a response he found strangely unset-tling. Bridges that were initially closed were reopened, and people evacuating from Manhattan by these bridges took much of the pres-sure off the waterside exodus. By 1700 on 9/11, Day and Wilson had assisted the maritime community in evacuating from Lower Manhattan a population estimated at more than 750,000 people.

The issues that confronted the Coast Guard team in the days ahead were as varied as could possibly be imagined. Besides the initial oversight of the evacuation, there were authorizations for on-water fuel transfers, orders for fuel and crane barges without pur-chase orders, the movement of multi-million dollar yachts from North Cove in absence of their owners, the development of protocols for inspecting vessels that had volunteered to help, and, for the first four days, expectations of a waterborne evacuation of thousands of casualties—an eventuality that faded with the growing knowledge that there would be few survivors.

At North Cove Marina, Lieutenant Day asked an officer from the buoy tender *Katherine Walker* to attend a meeting of emergency response personnel. The Coast Guard was told to "do whatever you think is the right thing to do to help." In the hours and days that followed, a few hundred feet from Ground Zero, Lieutenant Day and Chief Wilson directed more than seventy volunteer vessels in the transport of rescue workers and several tons of relief supplies from Jersey City. Within the first forty-eight hours, corporations began to contribute pallets of food, water, and clothing. These relief supplies were brought across the Hudson in such quantities that North Cove Marina became a depot. At one point, water traffic was so thick that Lieutenant Day became a human Vessel Traffic Service, physically waving vessels in and out of the cove. Eventually, using the local knowledge of the harbor pilots, two additional depots were established; one at the Battery and another at Chelsea piers.

When firefighters were unable to get fuel to fire pump trucks, water cooler bottles were filled with diesel fuel from the *New York* and a half-mile-long bucket brigade created. With building facades coming down around them as late as Thursday afternoon, the two representatives from the Coast Guard worked with the New York Police Department (NYPD), the Harbor Pilots, and the tugs, to develop a mass evacuation plan, a plan that was exercised three times that night and Friday morning. At one point, Lieutenant Day walked toward Ground Zero, just to get off the pilot boat for a few moments of relief. Instead, he came across a foot in a shoe, and remembered that he just stood there, fixating on it, for several moments. At another point, he later remembered seeing pallets of supplies being dropped off at North Cove, including pallets that held 20,000 body bags for the anticipated casualties. "It was like a war," Day recalled thinking. "Like a siege."

On board *Hawser*, Chief Todd's primary concern for his crew—apart from the general level of debris in the air—was whether they were protected from any possible biological or chemical agents that terrorists might have released into the air. Todd, who sensed early on that the country was under attack, was not one to take chances when it came to the defense of his cutter. He had his crew outfitted in Oxygen Breathing Apparatus (OBA) and full firefighting kit. Most of his crew was from the New York area, and they were in disbelief that the landmarks that they had grown up with were being destroyed. But, Todd noted, the general public still had not made the connection between the collapse of the towers and the fact that the country was now at war.

> One thing that was very strange about this: you see that the buildings have collapsed, you see people choking, and you pull up to the bulkhead, and they're asking you 'where're you going?' They want to know 'you going to Hoboken or are you going here or there?' And we're telling them, 'Sir, we're going anywhere but here. We're getting you off the island.' So they would wait for another tug that might be going to their specific destination.

For the people who did accept escape on board *Hawser*, the crew gave them fresh water and fashioned gas masks for them. The tug maintained a security zone around the area of the Trade Center, ran food supplies to *Katherine Walker*, and acted as On Scene Commander for several bomb threats to area bridges ranging from the Brooklyn to the Tappan Zee. It reinforced Chief Todd's decision on 9/11 to break out his entire weapons suite, such as it was, consisting of a quartet of 9mm pistols and two shotguns. "You gotta go with what you got," said the Chief wistfully, who despite the meager weaponry had always maintained his crew and his cutter as a professional military platform.

Lieutenant Christopher Randolph, a 1995 Officer Candidate School (OCS) graduate, was on board *Adak* on the morning of 9/11, familiarizing himself with the 110 in preparation for assuming command of another 110 berthed at Sandy Hook, the *Bainbridge Island* (WPB 1333). After three enlisted tours and three officer tours on board high endurance cutters, Randolph's 'fam cruise' on *Adak* started on September 9th and was supposed to last about three days. In fact, September 9th was the first time he had ever seen New York Harbor, and he remembered how he had spent much of that day looking up at the enormous twin towers. He remained with the *Adak* for more than two weeks.

Ironically, Lieutenant Randolph had just returned from a tour of duty in the Middle East, as part of the Coast Guard's mission as maritime interdiction coordinator for Iraq. Because of that experience, he immediately thought of terrorism on 9/11. During his time on *Adak*, Randolph saw how Lieutenant MacKenzie placed great responsibility on the shoulders of his petty officers, in particular his QM1, Matt James. When he assumed his first command, two weeks later on *Bainbridge Island*, Randolph carried with him many of the lessons he learned that morning. The first thing he did was to go out and find a 'battle ensign,' an overly large American flag which *Bainbridge Island* flew continually as it patrolled New York Harbor. "It was great for the morale of the crew, as well," Randolph said. "People would be walking up and down the East River seeing us and screaming and howling and honking horns. When you pass under the Brooklyn Bridge and everybody starts honking their horn at you, you know you're doing good stuff."

When Admiral Bennis saw the scope of the devastation, he put out a call to the rest of the Coast Guard to send everything. He asked for everything he could get his hands on. Communications were down everywhere, with sporadic cell phone channels getting through, and Bennis improvised every technique he could think of to maintain contact with Captain Harris at Fort Wadsworth.

At one point, he made contact with his son and asked him to go to Activities New York and tell Captain Harris to call Sandy Hook, where land lines had not been cut, and Bennis would speak with him that way. When an M-16-toting petty officer stopped Bennis's son at the gate, he called his father, who asked him to hand the cell phone to the petty officer so that the admiral, who knew the guard by his first name, could clear him through. The lack of communications led to continuous frustration, as those outside the New York area sought out Bennis for situational awareness.

> I would take my little bitty cell phone, and I'd take it outside the command suite at Activities New York, and I'd lean against the bicycle rack near the galley, because that was where I had the best reception. I had a button that said "Loy," which I'd programmed in, a button that said "Allen," which I'd programmed in, and a button that said "GNN," for George Naccara in Boston. And I'd call them on this tiny cell phone and give them the status report, telling them what we were doing, and that I would call

them in forty minutes, an hour, and hour and a half. And I said to the Commandant, "You need to know that everything we need to be doing, we are doing. We're doing it well. But we need more people, and more logistics support. But everything you think the Coast Guard should be doing, or needs to be doing, we are doing it." The only thing we're not doing, is demonstrating it to those of you outside of New York.

If there was one glaring deficiency in the aftermath of the 9/11 attacks, it was in communications. Admiral Bennis could communicate by radio with the units within Activities New York, but contact with the world outside was severely limited. When he did reach the various Coast Guard offices outside of New York over the following hours and days, the admiral would be subjected to oftimes inane questions during the ensuing phone conferences. Someone would ask if he had seen the latest security notice. Bennis would remind them that Activities had no phones, no computer connections. But the notice was faxed to your office. You don't understand, Bennis would say, faxes are phones, and we have no communications; all lines are down.

Communications did not begin to come back on line until a communications trailer from Atlantic Area's Communication Area Master Station Atlantic (CAMSLANT) arrived from Virginia and was set up on the point overlooking Fort Wadsworth. The Atlantic Strike Team's Mobile Incident Command Post (MICP) arrived at Bayonne to further bolster the shaky communications net. Maintenance and Logistics Command (MLC) Atlantic sent a hundred and fifty cell phones. As each of these elements arrived on scene, communications gradually improved.

Eventually, the briefing from Bennis over his cell phone came to be held at 1600 each afternoon. Wherever Bennis happened to be at that moment, he would call the Commandant. These phone conferences were joined by the Area and district commanders, the Commander of MLC, Rear Admiral Bert Kinghorn, and all would be on the other end of Bennis' phone call.   The Commandant would welcome everyone, then turn the line over to Bennis, for a view from Ground Zero. On the evening of 9/11, as he reported that an ice rink was being prepared as a morgue for casualties whose numbers were estimated in the tens of thousands, he heard nothing but silence from his fellow flag officers at their stations around the country.

Even two months later, as the site continued to burn, Bennis kept up his daily brief to the rest of the Service. The response from around the Coast Guard focused on the need to get telecommunications specialists into the area to reconnect Activities New York to the outside world. Phone lines on Staten Island were eventually rerouted through Denver, and when Bennis got his new phone and called his wife, she told him that the caller ID read "Tony's Pizza." Don't ask why, Bennis told her. It works.

The crisis showed the value of personal relationships amongst the principles as well. Bennis considered Rear Admiral George Naccara his best friend. Bennis had known Admiral Loy for many years and had an excellent working relationship with him. It was not a situation where he had to introduce himself to any of the chain of command, and the senior leadership in turn was confident in the abilities of Bennis and the capabilities of Activities New York. As the crisis began, Bennis looked to the senior leadership principally to find out "what was in their heads."

> I knew what I wanted to do. I knew from working with the city the best way to accomplish it. But what I wanted to know was, was I in fact a free agent. And I was. As the Commandant said later, he allowed his field commanders the freedom to let their creative juices flow, and I was able to do that. And at the start I needed to know if I was going to have that freedom or were they going to micro-manage me. They didn't send in a team to oversee the operation. Instead they sent exactly what I needed, the people I needed. It worked out very well.

Such relationships had been built, prior to 9/11, between the Coast Guard and the myriad law enforcement agencies throughout the tri-state area. Commander Michael McAllister, the Chief of the Marine Response Division at Activities New York and a long-time patrol boat skipper, was attending a course at the Coast Guard Academy in New London, Connecticut, on 9/11. Racing back to Staten Island on a largely deserted I-95, he needed only show his Coast Guard identification card to pass through the many police road blocks set up to control traffic around New York.

"This was a devastating incident," said Commander McAllister, "but it really did highlight the Coast Guard's role in response in the New York area. Nothing has had a greater impact on awareness of the Coast Guard's importance and actions in this port, both amongst the law enforcement community and the general public."

Even with communications shot, Admiral Bennis found himself tracked down very quickly. When the Port of New York/New Jersey closed, twelve commercial vessels were already waiting in a queue to enter the harbor. Within twenty hours of closing the massive petroleum port, calls were coming in from the White House with the none too subtle information that people were running out of gasoline in Portland, Maine. Fuel from New York supplied Logan Airport and dozens of other strategic locations up and down the Atlantic Seaboard. Reopening the Port of New York became a matter of intense importance, but before that could happen the Coast Guard had to secure it from maritime threats.

The largest concern for Admiral Bennis in the first twenty-four hours, after the initial evacuation of Manhattan, was with the possibility of more attacks, where they might come and how could he prevent them. Captain Harris likewise had no intention of letting anything into the harbor that hadn't been searched and declared safe. These concerns led to a series of evaluations: what had the terrorists done on 9/11 and, more importantly, what they had failed to do that they might seek to finish with a second or third wave of attacks.

With limited resources, the Coast Guard in New York was forced to rank a series of equally unpleasant scenarios. Prior to 9/11, the command had identified approximately 150 key assets within the port area. Theoretically, all had to be protected, yet given limits on platforms and personnel, hard decisions had to be made over priorities. The destruction of a single strategic bridge would close the entire harbor. Targeting the Indian Point nuclear power plant would spread both radiation and fear. Destroying an American symbol like the Statue of Liberty would both enrage and demoralize the population.   Then there were all the elements of a free-trading society that made the Port of New York and New Jersey a national asset and the economic nerve center of the East Coast: oil and chemical refineries and storage tanks in the Kills, as well as container ships and crane facilities, tunnels, towers, railheads, power grids.

In addition to the limits on how much port security it could provide, the Coast Guard was not in a position to board a terrorist vessel and thwart a ship's crew intent on mass destruction. This realization led to some quick improvisation.

"Against somebody who is intent on not allowing us to board, we wouldn't have a good way to do it," said Captain Harris. "Unless we assaulted from aircraft or disabled the vessel. The best thing we thought we could do was keep tugs near vessels, so if it looked like they were doing something that was unexpected we could deflect their aim."

If necessary, the 270' or the 110's patrolling the harbor could go after rudders. Closing the Kills waterway that separated Staten Island from New Jersey to recreational traffic and stationing a cutter at each end, as was done, allowed Activities to consolidate many potential targets under a single maritime security force lay-down.

The wartime feeling of the harbor was felt keenly by the Coast Guard personnel who responded. Marine Science Technician First Class (MST1) John Kapsimalis of the Atlantic Strike Team arrived at Battery Park after a trip across New York Harbor on the Staten Island Ferry. He remembered the absence of boat traffic in the whole harbor. "The Coast Guard had completely shut it down. You had cutters, you had raider boats, you had 41s. It was a military presence but a Coast Guard presense, and that just jumped right out: the Coast Guard just shut this harbor. You knew we were at war when you saw that."

If there was any guiding light that emerged from the eventual identification and analysis of the adversary, it was that they seemed to prefer the attack that generated the highest dramatic effect. This meant the most bodies on the television news, with destruction of the economy as a secondary by-product. For Admiral Bennis, this was classic terrorism, instill fear and destroy confidence. Given that, one could make plans. One did not need to understand the motives of mass murderers to find where they lurked, how they operated, and devise ways to stop them. Based on the highest dramatic effect quotient, his "gut reaction" told Captain Harris that the Statue of Liberty required a security zone around it immediately.

For Bennis, from the first moment he arrived at Sandy Hook, he knew that those underneath him looked to him for leadership. He had a team in whom he retained complete trust, and he let them know that right up front. They would work this problem together, without burning each other out. During the endless rounds of briefings that followed the attacks, he would only ask a probing question here or there, just to make sure that everyone was headed in the right direction. Bennis was the opposite of a micromanager, and believed instead in empowerment, in telling his force to 'go make magic and be brilliant.' After 9/11, he was not disappointed. "We built the command together, and if didn't work, we'd take it apart together and find a different way."

After the initial rush of adrenaline-charged days were over, Bennis circulated amongst his team to find those personnel who needed to get away, needed family time, and prodded them away from their posts, even for an hour.   Bennis had been involved in Coast Guard responses to major disasters, notably the *Exxon Valdez*, but for him they all paled in comparison to 9/11, and for a simple reason. There was nothing in those other events approaching the scale of human horror produced by 9/11. A prudent commander remained vigilant for signs of shock amongst his people.

The concern went both ways. At one point in the crisis, Bennis heard a petty officer talking with one of his chiefs. He didn't know what made him do it, but Bennis paused out of sight to listen in. The petty officer said, 'Chief, we need a break; we need a couple hours off.' To which the Chief responded, 'Goddammit, the Old Man just had brain surgery and he's been working twenty-three hours a day for five days!' Bennis smiled to himself, then tiptoed back in the other direction. It was at such moments that the Admiral allowed himself to believe that perhaps he had been kept around for a reason.

In terms of concepts of operations, based on their experience during OpSail 2000, much of what transpired on 9/11 had been forseen: shoreside security, securing bridges, tunnels, power plants, and so forth. One area that had not been forseen, and perhaps could not have been, was the need for the sudden and catastrophic evacuation of a large section of the island of Manhattan. And that evacuation was hardly a controlled one. Much of the population that fled to the waterfront was, for good reason, terrorized. A new van that belonged to the command had its side panel destroyed by a vehicle driven by someone who sideswiped their way out of Manhattan in their desperation to escape.

Bennis and staff discussed at some length whether or not it would have made a qualitative difference in the Coast Guard's response if the Service still operated from Governor's Island. In the end, they decided that the island's tactical and strategic advantages would have been outweighed by other concerns. Amongst these would have been the necessity of family evacuations, the require-ment for personal protective suits in the midst of the poisonous cloud that drifted over the island for six weeks, and the knowledge that their command post was set up, in effect, in a hazardous waste site.

Given that Fort Wadsworth was the right place to stage the Coast Guard response, Bennis saw other areas that could be fixed. Maintenance and Logistics Command Atlantic and its commander Admiral Kinghorn came to the rescue with communications, but the Service required more port security units, and needed to be able to keep them on scene longer. Two 38' Deployable Pursuit Boats (DPB), the Coast Guard's 'go-fast' boats for chasing drug running boats, from the Tactical Law Enforcement Team (TACLET North) in Yorktown, Virginia, which in recent years had been mothballed, found their element in New York Harbor.

"Without knowing it, that is what those boats were designed to do," believed Bennis. "That is where they were designed to be.

They were a very good tool for a public that respects and appreciates the Coast Guard while at the same time believing that the Coast Guard can't catch a Boston Whaler. We got the Deployable Pursuit Boats early on, and I would have them run from the George Washington Bridge to the Verrazzano Narrows Bridge at speed several times a day, just so people could go, 'What the hell was that? It's the *Coast Guard.*' Just knowing we had that capability gave a lot of people pause, be they tourists or people intent on violating a security zone."

The Coast Guard offered comfort as well as pause. Awaiting a visit by the President, Bennis found himself talking with a firefighter from Ground Zero whose company had lost sixteen men that day. "He said, 'You know how you feel when a fire truck comes through the neighborhood and you feel reassured. That's the way my family feels when they see the racing stripe out on the water."

It was a feeling Bennis knew well. He had made a conscious decision to recommend against the deployment of Department of Defense (DoD) assets to New York Harbor. He wanted the battle group that was offshore providing air cover to stay offshore. While the racing stripe offered reassurance to the public, DoD platforms, in Bennis' view, served as an unsettling reminder that another attack might be imminent.

> Rather than the one or two racing stripes the public was used to seeing on the water, they now saw forty-five or fifty. The public is used to seeing us work in tandem with the NYPD, so the blue boats with their white stripes are often alongside the white boats with the orange stripes. That being said, we made sure that the media carried as many pictures as we could get of our gray-hulled, black-striped PSU boats, to let them know that this was a different operation. We never deploy those folks domestically, but now we did, so people knew we had a heightened level of security, beyond what the Coast Guard usually has on patrol.

Bennis also led several discussions on what the Coast Guard was going to show, versus what it was prepared to use. A .50 caliber round fired from one of the dozen 25' Transportable Port Security Boats (TPSB) in the harbor was a threat to anyone within four miles. Rules of force and engagement were laid out early on, and the PSUs adapted accordingly. With the arrival of the PSUs, and given the hardening of potential targets by the Coast Guard, the Port Authority Police, NYPD, and New York and New Jersey State Police, Bennis believed that a second wave of attacks, if it came within the maritime domain, would not have proceeded very far.

Commander McAllister experienced firsthand the difference in public response to Coast Guard platforms painted white or gray. It was about five days into the crisis, just after Port Security Unit 305 had deployed its gray boats to protect the U.S. Navy hospital ship *Comfort.* After attending a meeting of the New York mayor's office of emergency management, McAllister was mobbed by the press. They wanted to know what kinds of guns the PSU boats had, where they came from, were they Navy or Coast Guard boats, and who were they out there to shoot and who to protect?

"There was definitely a sensitivity to our overt combat readiness," remembered McAllister. "I think if gray hulls had been in the harbor, it would have raised everybody's awareness or concern to a much higher level. The Coast Guard puts people much more at ease, and we're far more capable of identifying threats to safe commerce and safe recreational boating than the Navy is. And that helped us transition much faster to the 'new normalcy.'

For Bennis, the challenge ahead for the Coast Guard was to secure not only U.S. but foreign ports, to secure container cargo from point of origin to point of destination. With his cancer temporarily at bay, he had new challenges of his own. Before he had the chance to enjoy a formal retirement ceremony, Bennis was tapped to join the new Transportation Security Administration. He was followed there by Admiral Loy. Once again they were working together to develop secure modes of transport for both people and cargo.

Only now, they had some time and the phones worked. Time and communications. From 9/11, they understood that in any future crisis, those were the two elements likeliest to be in short supply.

Chapter Two

The Atlantic Strike Team Moves Out: Maritime Safety at Ground Zero and Fresh Kills

In early August, 2001, Staff Officer Ellen Vorhees, a member of the Coast Guard Auxiliary, had just started to help out at the front office of the Atlantic Strike Team (AST) located at Fort Dix, New Jersey. She came in once a week, on Tuesdays. September 11 was a Tuesday, and for Vorhees it was a normal, busy morning at the unit. New members of all three strike teams were at Fort Dix for an indoctrination course known as the National Strike Force Boot Camp.

When a call came from the wife of the Strike Team's Operations Officer, Lieutenant Scott Linsky, that a plane had hit one of the World Trade Center towers, Vorhees and a few others went into the conference room to see the event on television. It seemed like an accident, so she went back to her post at the front desk, just as the second plane hit. Later that afternoon, Fort Dix was closed, and Vorhees was sent home. Vorhees was old enough to remember Pearl Harbor, and how frightened all the adults had been. She now felt the same way. She was able to return to Fort Dix the following Tuesday, by which time the headquarters of the Atlantic Strike Team was quiet. Almost the entire unit had long since gone to Ground Zero.

*   *   *

On 9/11, Chief Petty Officer Dan Dugery had served as a

marine science technician in the Coast Guard for more than a quarter century. Earlier in his career he had served as a marine science instructor at the Coast Guard Academy. Some of his cadet students were now in command of cutters. For his latest tour, he had served for several years in the cavernous, spotless former aircraft hangar with its fifteen-ton overhead crane at Fort Dix that served as staging point for the Atlantic Strike Team.

Aircraft pallets with anti-pollution gear, oil skimmers, pumps, oil booms, jacob's ladders, inflatable boats, air monitoring equipment, chemical suits and weapons of mass destruction kits, were battened down and ready to be driven by flatbed or air-lifted by C-130 to any environmental emergency. All-terrain vehicles were rigged to conduct riverbank survey work amongst the debris and pollutants left behind by floods. A 32' Sea Ark multi-purpose vessel powered with twin 225 hp Evinrude outboards was fitted with a flat bow that could be lowered to deliver pumps and oil booms onto beaches like a tank landing craft. By rigging a series of outriggers, booms, and pumps, a Vessel of Opportunity Oil Skimming System (VOSS) could convert any vessel 140' or larger into an oil-skimming ship. Two 48' inflatable rubber barges designed with huge hull pockets could each handle 22,000 gallons of recovered oil. The Team had even used its 22' jon boat for search and rescue work during flooding in southern New Jersey, at one point grounding the vessel on the roof of a submerged van in a flooded parking lot. All of the different Coast Guard Strike Teams (Atlantic, Pacific, and Gulf), were fitted out with identical suites of gear, and operated on the assumption that they would receive no outside support wherever they were sent.

With tons of such environmental response gear cleaned, prepped, and ready to deploy, Dugery had seen the AST load up their flatbeds and move out of the hangar in less than forty-five minutes, a turn and burn speed that came only with constant drilling, training, and practice. In recent years, the AST's services had been required so often that real-life emergencies often took the place of practice.

Dugery and the unit's other chiefs had gathered on the morning of 9/11 in the conference room off the hangar. Several members of the unit had made E-7, and the chiefs were discussing the particulars of the upcoming Chief's Call to Initiation, when Lieutenant Linsky came into the room and said, "Turn on the TV. Something's happened in New York."

*   *   *

Lieutenant Commander Nathan Knapp, with tours on the high endurance cutter *Mellon* (WHEC 717) and at MSO Cleveland and MSO Hampton Roads, arrived as Executive Officer of the Atlantic Strike Team at Fort Dix in June of 2001, with years of port operations experience. Knapp was bouncing between his duties in the ongoing boot camp course and as XO when the televisions throughout the building began tuning to CNN. Within a few minutes of the second plane going in, Knapp called his commanding officer, Commander Gail Kulisch, who was on the road, and the two went over how they might be called upon to respond.

With the buildings still up, Knapp as yet saw no role for the unit. The Atlantic Strike Team was a Headquarters unit, generally deployed in response to a call by a captain of a port for environmental response work. Such calls could come down to Fort Dix along many paths, but the result was always the same: instant mobilization. When the first tower started to fall, Knapp recalls, "everybody wanted to get into the fight." The televisions were quickly muted. Everybody knew it was time to get to work.

As one of the Coast Guard's front-line response units, all knew the call would come, and soon. With Lieutenant Linsky, Commander Knapp surveyed who was on board and who they would need to call in. Chief Dugery was just getting set to make phone calls, when he saw the people he was going to call walk through the door. Reserves like MST1 John Kapsimalis had headed for the unit as soon as they learned of the news.

Unlike most Coast Guard shore units, which are split into several roughly equally-sized divisions, at the Atlantic Strike Team the Operations Division (Ops) was essentially the entire unit. With the exception of a few Administration staff, everybody worked for Ops. When a request for assistance came in, whether from the Environmental Protection Agency (EPA), the Federal Emergency Management Agency (FEMA), or from the Coast Guard itself, it was Lieutenant Linsky, Ops, who put together the response team.

Linsky found his Warrant Boatswain, Leo Deon, and told him to double the size of the watch section. He found his senior truck driver, Chief Damage Controlman Bernard Johnson, and told the Chief to get three drivers, and equipment bags, prep the mobile command post, and get ready to go to New York. "It's pretty common," said Lieutenant Linsky, "that we will see things on CNN and say, 'We're going to be there in twelve hours.' Even before the towers collapsed, I had a pretty good idea that Activities New York was going to want us there to set up the Incident Command System. Once the towers collapsed, I was pretty certain that the EPA was going to bring us in on the hazardous materials side."

As the second tower fell, the phone rang; it was Commander Kulisch. Like Knapp an experienced port operations and marine safety response officer, she wanted the Strike Team's boats, coxswains, and engineers deployed for what she knew would be immediate SAR and port security work. Being on the road near Staten Island, she was also in a position to alert the boat crews leaving Fort Dix of where highways were closed, or where they were clogged. The Strike Team's 32' Sea Ark, as well as its 23' boat, were rapidly loaded onto a flatbed and left for Station Sandy Hook. Sandy Hook called to report that landline communication with New York was gone, and that a radio message from Admiral Bennis had requested the Strike Team's mobile command post.

Activities New York also asked the Strike Team to set up a staging area at the Military Ocean Terminal at Bayonne, New Jersey (MOTBY), to provide support for the Coast Guard's small boats then gathering by the dozens in New York Harbor. Knapp ordered the Strike Team's mobile command post, which Ops had already readied for travel, to Bayonne. Linsky told Chief Johnson, the Strike Team's senior commercial driver that the Chief was now Team Leader and it was his decision, once on scene, where the command post should go and how it should be set up most effectively.

The Mobile Incident Command Post (MICP), deployed on the back of a tractor trailer truck, consisted of two sides, one for admin and the other for comms, each with its own generator, heating, and cooling. Each could be put on the road separately, if need be. A sleeve fitted to each side created a conference room and more office space. To be able to operate in an area where power was down, the whole 40,000 pound command post was cranked up and put together manually, without any hydraulics or electronics, usually by six people in three hours.

The Atlantic Strike Team's MICP was out the door by 1105, and was later joined by a similar unit owned by the Gulf Strike Team. Together, they formed a staging area for food, water, fuel, crew rest, and communications with Activities New York, for the small boats of the strike teams, as well as local Activities small boats, and, later, the armed Boston Whalers of Port Security Unit 305. "They [Activities New York] basically pulled their OpSail 2000 plan off the shelf," said Lieutenant Linsky. "Because we supported OpSail 2000 with the Mobile Incident Command Post at MOTBY."

As the Threat Condition at Fort Dix was raised, and raised again, Chief Dugery led a squad to secure the perimeter of the Strike Team's facility. He double-checked the generator in case the unit lost power. He remembered, as he walked the perimeter, seeing the Military Police nearby with blue magazines in their rifles, indicating non-live ammunition. As Dugery watched, a Humvee rolled up, a Master Sergeant got out and spoke to the MPs, and the blue magazines instantly came out and live magazines went in. It began to hit home that things were not normal anymore.

Linsky realized after talking with Sandy Hook that this operation was going to be different than a typical port security mission. As casualty figures out of New York rose throughout the day, so did the likelihood of a massive medical evacuation from Manhattan. "Before the boats went out the door," Linsky remembered, "I structured the boat crews a little bit different. I sent a coxswain, an engineer, and a boat crewman, but I also sent an EMT with each boat crew, assuming that we were going to be responsible for some medevacs."

Boatswain's Mate First Class David Bittle was the coxswain for the Strike Team's 32' Sea Ark as it was launched from Station Sandy Hook, on the afternoon of the 11th. Chopped to Activities New York, Bittle drove the 32 foot boat from Sandy Hook to Station New York on Staten Island. Small boats from around the Coast Guard had converged on Station New York's small dock area. Bittle's Sea Ark was chopped again, to the cutter *Adak*, which was acting as on-scene commander pending the arrival of the cutter *Tahoma*. The 32-footer maneuvered into North Cove Marina to help there, the start of a ten-hour patrol that would not end until the next morning at 0800. On the 12th, the small boats set up a twelve-hour on/off patrol schedule. For the Sea Ark, this meant overnight patrols from 1900-0700. Bittle remembered the crowded scene at Station New York, with the traffic eventually so bad that the Sea Ark was moved from Staten Island to the MICP staging area at Bayonne. Through the MICP, the thirty-two was able to communicate with both Station and Activities New York.

As the emergency grew, and his crews dispersed around New York, Lieutenant Linsky started dipping into the basic training course for personnel. He sent these off to New York. Very quickly, he had more than ninety Strike Team personnel from around the country in the field. Chief Dugery had gone home in the afternoon just long enough to tell his wife that he'd see her in a few weeks. Then he was gone. At 2:00 a.m. on 9/12 he arrived at Activities New York. Driving from Fort Dix to Staten Island he looked at the reddish-orange glow from the fires at Ground Zero. After the New Jersey Turnpike exit near Edison, it suddenly occurred to him that there was no traffic on the road. A constant stream of lights on the harbor showed the massive fleet moving rescuers and supplies into the city.

Each of the three Strike Teams carried about three dozen active duty personnel, and Linsky eventually fielded nearly eighty-five per cent of them, almost the Coast Guard's entire National Strike Force, for the emergency in New York. Trained to the same standards, equipped with most of the same basic gear, familiar with trading personnel back and forth, depending on the scope of past environmental emergencies, Linsky was confident the three teams would integrate seamlessly. The operations officers of each team knew and had worked with each other before. In any response situation, they had the luxury of calling one another directly. This allowed them to slice through what otherwise would be a knot of bureaucracy when time was of the essence.

Lieutenant Commander Knapp moved up to Activities New York to integrate into the planning staff there. Commander Kulisch was already there, where she noted that everyone on scene, from Captain Harris on down, had defaulted into their Incident Command System mode of operation, which involved continual assessment of the problem and then determining how to work it through. Knapp had worked for Rear Admiral Bennis during his time at Hampton Roads and now he would work for him again. He observed that Bennis managed to remain calm and to convey both stoicism and humor to his staff, even given the enormous strain of the crisis. At one point, while escorting First District Commander George Naccara through Activities New York, Bennis pointed out Knapp and remarked, "Admiral Naccara, have I introduced you to my illegitimate son Nathan Knapp?"

For Knapp, a competent leader that recognized when to push and when to pull up on his highly trained people helped maintain operational effectiveness, and such effectiveness was the key to the whole response. "That's what it's all about, what we're here to do. When everybody else in the general public was going to chaos, it was time for us to go to work. In a situation like this, you don't have time to be everywhere you would like to be, and you can't be wondering if your people are doing the right thing. You have to know that it's embedded, and that gets put in by training and good leadership."

The rest of the Atlantic Strike Team arrived in New York on the night of 9/11 to monitor air quality and worker safety as rescue operations began at Ground Zero.

> Within the first couple of days, between the Environmental Protection Agency and the Captain of the Port, they agreed and determined that the EPA would take responsibility for all hazardous materials-related issues

at the World Trade Center. And that would free up Admiral Bennis and his staff to focus on port security, after the initial medical evacuations. Once that happened, EPA Region II asked the National Strike Force to stand up a large response organization, which they did not have experience with doing. So they asked us to bring in our full force to create the Incident Command System, staff the main positions, and help them integrate into and give them that structure to work within. To the credit of their leadership, they recognized that, and said, 'We know we don't have experience doing this and the National Strike Force does.'

Starting on 9/12, a platoon from the Strike Team was responsible for air monitoring the financial buildings in Manhattan, making certain that rescue workers could go in and search for critical data and documents. Two days later, Chief Dugery joined and led teams that entered buildings in the financial district sliced in half by the collapse of the World Trade Center. These were buildings forty-to-fifty stories high with no electricity. As he entered one structure to test the air, he felt his work boot scuff on something in the debris pile and looked down to see a human forearm. "You shook off your feelings like a chill," he said. "Once we had ensured the safety of the air, we escorted representatives of these financial companies into their work spaces."

At one point, Dugery found himself in a secure vault in a bank basement, filling a rolling trash can with templates for the printing of gold and silver certificates.   "It was vital that these banks retrieve their records, computer discs, and other materials that would get their operations back on line as soon as possible. We needed to get the stock market and the economy back on a stable footing." As he made his way through the dust and debris of lower Manhattan, he was struck by the irony of his situation: a Coast Guard Chief pushing a trash can holding one of the keys to the recovery of an entire nation's economy. MST1 Kapsimalis remembered filling a letter cart with trash bags full of blank bearer bonds stacked three feet deep. "We were under tremendous time pressure to get these bonds out, both because of their importance to the national economy, and because of the instability of the buildings. I asked the bankers I was with, 'Can I assume we're circumventing some procedures of yours?' and they all looked at me and said 'You wouldn't believe it.' That's how surreal it was."

For other members of the Strike Team, their response gave them a unique view of the crisis. Being charged with air monitoring in some of the biggest buildings in the world, they looked down from their work on long bucket lines of men and women, on dogs crawling over the pile and digging through the rubble.

MST1 Robert Schrader, a Coast Guard Reserve, escorted a group of corporate officers to the 44th floor of an office building on Broadway, to gather data from their PCs and servers, when a laser alarm went off signaling that the building was flexing and had become unstable. Schrader looked down to see people running out into the street more than 400' below. "There was quite a 'pucker factor' there," he recalled, so everyone was gathered up and escorted back down to the foyer and out onto the street. Checking the building at 64 Wall Street for carbon monoxide, Schrader remembered that his group of Coast Guard personnel were the only ones there. "We were all alone. Wall Street was empty. Broadway was empty. We could go into any building we wanted to, and anywhere in that building. The city was stopped, and it was like walking through the movie *The Day the Earth Stood Still*. Ground Zero was one block away. There was no electricity, no sound except for sirens in the background. When it rained that Friday it turned the dust into oatmeal, and when it dried it turned to paper mache."

At the Merrill-Lynch Building at 222 Broadway, Chief Dugery found that two sides of the building had debris splashed up against them, "like sand on a beach," from the pulverized trade center buildings. The other two sides of the building had so many vehicles arrayed alongside them that entry was impossible. Where they could enter the massive financial structures, his teams sampled the levels of oxygen and carbon monoxide, while monitoring the air for hydrogen sulfide, a substance which both deadens the sense of smell and then kills personnel who mistakenly enter pockets of the lethal gas.

At 2 World Financial Center, teams hauled their gear up more than twenty-five floors, then worked their way down through office spaces filled with gray dust. All rescue workers were covered in the same gray dust, as if the entire city had been turned into a black and white photograph. On one conference table the Chief noted several cups of coffee, and underneath the table a pair of women's shoes, "as if everyone had just got up and went to lunch, or that they were expected back shortly. Then you'd walk to the other side of the building and everything would be completely wiped out."

MST2 Monica Allison remembered standing in a staging area, waiting for orders, when a member of a visiting fire department pointed out a set of fireman's turn-out gear up in a tree. For Chief Warrant Officer Leonard Rich, the crisis hit home as he walked through the debris and saw the badge of a fireman's helmet. He asked and was assured by the Federal Bureau of Investigation (FBI) agent with him that it would eventually get to the family. BM1 Patrick McNeilly lost a high school classmate, a firefighter killed at the World Trade Center. Yeoman (YN) First Class Matthew Leahy, born across the river in Jersey City and raised a bit north in Guttenburg, lost a childhood friend, a Port Authority Police lieutenant named Robert Cirri whose body was found at the base of Tower One more than five months after 9/11. "It was tough," remembered Leahy, who worked at Ground Zero for more than six months. "I had a tough time controlling my emotions. It was a rollercoaster, from sadness one moment to anger the next. Every time I looked at the debris field I got angrier and angrier and angrier."

A week after 9/11, Chief Dugery was at Ground Zero itself, with other Strike Team members, monitoring the air quality on the massive debris pile to try to ensure the safety of the workers there. The list of hazardous materials produced when the two trade center towers crumbled was almost overwhelming. There were varying amounts of asbestos, PCBs, batteries, compressed gases, paint thinners, photo chemicals, fuel oil, gasoline, copier chemicals, medical waste, compressed gases, drugs, laboratories, ammunition, plastics, foams, and all the hazards associated with combustion, chlorine compounds and hydrogen cyanide. A blood bank was located in World Trade Center Five. Each tower held 160,000 pounds of R22 refrigerant for the air conditioning systems, which in confined spaces would displace oxygen. Heated above 1300 degrees F, it forms phosgene, a highly toxic and corrosive gas that causes fatal respiratory damage if inhaled. Then there was the aerosolized blood from the as yet unknown thousands of human beings killed as the structures crumbled.

"We talked for days at the beginning [about site safety concerns]," said Lieutenant Linsky. "How much asbestos was actually in the air? Thousands of gallons of dry cleaning fluid were burning. There was gasoline and anti-freeze and motor oil and office furniture on fire. It reminded me of going to a metal plating plant fire or an industrial park fire where you spend the first two days just trying to figure out what hazards you're up against." For Chief Dugery, it was a matter of tripping down a well-established list. "You start with the basics. Is there enough oxygen? Are there explosives in the air? If you get past those, then you start dealing with specifics, like air-borne particulates."

Chief Warrant Officer Leo Deon, an operational boatswain by trade and now one of the National Strike Force's lead experts in oil spill response, noted that it was several days into the crisis before people started to consider the potential for long-term health issues that might derive from the environment at Ground Zero. Rescue and response personnel had plunged into the 'pile' at Ground Zero with little or no thought to personal protective gear. Deon was charged with setting up stations where people and vehicles could be cleaned after working on the pile.

Deon also had to locate strategic sites where truckloads of pile debris itself could be accessed for criminal evidence by FBI agents, before the debris was washed and removed to the Fresh Kills landfill site on Staten Island. There the debris would be subject to even more sorting, investigation, and decontamination. The sensitivities of those desperate to locate colleagues and loved ones at the site limited the Strike Team in their ability to promote worker safety, in what was to them, in every other sense, a hazardous waste site. As Warrant Officer Deon remembered:

Being charged with establishing these wash stations, and vehicle washing areas, we could not mandate that these rescue workers, whose shifts had just concluded, go through and decontaminate. We were not allowed to even use the word 'decontaminate,' which is the proper term for a site of this nature. So we had the wash stations available, yet they were not mandated for use. Through some entrepreneurship, we were able to entice people to the wash stations, with cold drinks, with rehab areas where they could actually sit down, and through this aggressive approach a visit to the wash station eventually caught on. When word got out that there were some nasty materials that you could potentially be inhaling or absorbing through your skin or left on your clothing, then it became very prudent for people to go through and wash down once they left. It took a long time to transition from a rescue and recovery site to a hazardous materials site.

"We're Coast Guardsmen first," said Lieutenant Commander Knapp. "We think about the safety of human life. That was everybody's primary concern, including ours, just to get in and save people. But unlike the majority of the Coast Guard, our unit is focused on the nasties, what are we going up against here, what could we be called in to do." One of the unit's biggest concerns was with well-meaning rescuers who were nevertheless putting themselves at great risk amongst the multifarious hazards they knew to exist at Ground Zero.

For Chief Dugery, amid the acrid smell of scorched dust, it meant repeated cautions to personnel to make sure they were wearing their respirators, goggles, and hearing protection. MST1 Kapsimalis remembered using simple paper masks for much of the first week, despite feeling that such masks were more or less useless. "After a week or so, everyone was issued a half-face mask. I got done listening to CNN saying that there were no elevated levels of asbestos. Really? I purposely didn't eat on the Spirit of New York because of it. And I missed that young lady from *My Cousin Vinny*, Marisa Tomei, who was out there handing out food. That's alright. I went looking for a Campbell's food truck."

By 20 September, Chief Warrant Officer Leonard Rich had written a site safety plan for Coast Guard and EPA personnel working at both Ground Zero and at the Fresh Kills Landfill site, but admitted that trying for a comprehensive site safety plan at either spot was "like a child wrestling a bull." At the Fresh Kills landfill site, a treeless mound 130' above sea level, Rich and his Strike Team personnel set up air monitoring and site safety protocols. There the City of New York was transporting not just debris from Ground Zero—there were separate piles for wreckage from Tower 1 and Tower 2, as well as the other collapsed Trade Center buildings—but crushed rescue vehicles, police cars, and fire trucks. Some destroyed vehicles were eventually tagged for preservation as part of future museum exhibits and memorials. A temporary morgue had been arrayed for the body parts being found and removed from the continuously-arriving debris. Dogs brought in to narrow the search for bodies were ultimately ineffective because the smell of death was everywhere.

Buses filled with personnel sorting the debris arrived at the site to see a Coast Guard flag flying over a trailer unit. Nearby, Rich and his team set up areas where workers could get respirators, gloves, whatever personal protective gear Rich and his safety officers felt they should be wearing. And there the workers would clean up after their shifts. Separate areas were established for washing down vehicles as they entered and exited the site. As soon as debris landed at the Fresh Kills site, Chief Petty Officer Robert Field and other members of the Team set up air monitors at the site, sampling for asbestos and other hazardous airborne materials.

Over time, Lieutenant Linsky became worried about the mental health of his personnel. Coordinating operations from Fort Dix, he was receiving calls around the clock from Team members who reported finding bodies, seeing bodies, and he realized the toll it was taking. He decided to go to Ground Zero himself and that is when he saw firsthand how many physical hazards his personnel were facing. "I've never been scared in this job until I had to climb up the stairwell of a fifty story building which the structural engineers thought 'should be safe.' All the structural engineers could give us was a guess." Commander Kulisch likewise rotated from a command and control to a tactical mode, traveling between Activities New York and Ground Zero, to participate with the Team in these surveys and double check on their safety.

On the other side, personnel received the thanks, and sometimes more, of the people in whose corporate spaces the Strike Team was operating. Lieutenant (junior grade) David Reinhard, the unit's Assistant Operations Officer, remembered that his teams were offered almost every kind of gift imaginable. "They tried to give us money, they tried to give us furniture, and one guy tried to give us a painting by Picasso. They kept telling us 'that is our way,' and we had to repeatedly explain that we couldn't accept any gifts." The team led one anonymous corporate manager back into his offices, chatting with him amiably. On the way out one of the team members noticed a framed cover of *Fortune Magazine* with the face of the man they had just been talking with. He was the owner of the corporation.

Ironies abounded. Strike Team members were public servants in a frugal Service, yet they were reopening private corporate spaces. One noted that he was walking up the same stairs with an executive about his same age, yet the executive probably earned one hundred times more than he did. One remembered that a picture of the World Trade Center was on the cover of FEMA's emergency response to terrorism course book, and wondered if they would now change covers. Strike Team members were used to showing up at industrial disasters to have people see their Coast Guard coveralls and ask them where their boats were. No one asked that question in New York. MST1 Robert Schrader was struck by the complete absence of racism, bigotry, and hatred. MST2 Allison remembered how friendly people were, with no barriers between them. It reached a point where Team members began to feel guilty for being treated as heroes for just doing their jobs.

Chief Warrant Officer Deon, a lifelong ship driver, had long experience with the maritime public, which wasn't always happy to see the Coast Guard show up in its law enforcement capacity. As a member of the Strike Team, he was accustomed to going into a disaster area, doing his job, and leaving in more or less complete anonymity. "I think the City of New York knows we were there, and what we were there for, and appreciated it."

*   *   *

Commander Kulisch had little time to ponder the magnitude of 9/11. In early October, anthrax spores began turning up in Washington, DC, and in Florida. Exhausted by three weeks of continuous physically and emotionally draining work at Activities New York and Ground Zero, she drove south to become Deputy Incident Commander on Capital Hill. As soon as the first reports surfaced, Kulisch knew that the unit would be responding immediately, either to hoaxes or to actual incidents. She had already gathered her response officers together and had them formulate policy and decide how the Team's capabilities matched up with the estimated threat. The Team knew within the first week of its response that the anthrax on Capital Hill had been "weaponized," made into a much more lethal, easily airborne, variety of killer.

"We were ramping up our weapons of mass destruction capabilities even before 9/11," said Chief Dugery. "We had some test kits. Unfortunately, for [a biological threat like] anthrax, you need a good lab, until the technology catches up [for field sampling]. If you're trying to identify the pathogen you need to grow it in a medium to find out what you have. It takes time. Chemicals will react instantly. If it turns something blue you know you have chemical 'X'."

Lieutenant (j.g.) Reinhard was initially happy to leave New York and go to Washington to lead the team's anthrax response. He saw the overwhelming kindness exhibited by both responders and citizens during the first weeks of the crisis returning to the impatience famously associated with New York. But in Washington he ran into the food chain that was the federal government, and discovered quickly that he was not considered very high on it. "I was under-gunned. I could do the job. I wasn't worried about my S.O.P. But I was there [at the Senate Office Building] on a Saturday, with the Sergeant at Arms and three Senators telling me that the building *would* be open on Tuesday. I kept telling them, 'there's no way.' Then they would respond, 'oh yes it will.' And they looked at me as some twenty-five-year-old kid. What could I do then? I couldn't explain that it was not safe to do that." Reinhard would have preferred to have been a Commander right then, and in fact he believed the situation was not properly understood by the politicians until Commander Kulisch arrived on scene and made it plain. By the end of the Strike Team's response, and with the knowledge of how narrowly they had escaped death, the politicians were listening to the Lieutenant (j.g.).

"I can't separate World Trade Center from the Capital Hill anthrax response," said Lieutenant Commander Knapp. "Together they were the *Exxon Valdez* of weapons of mass destruction (WMD). We had been talking about WMD before 9/11, but on our own time schedule. We no longer have that benefit."

Chapter Three

9/11 at First Coast Guard District in Boston: George Naccara and the Northeastern Maritime Frontier

Rear Admiral George Naccara was at his desk at First Coast Guard District headquarters on the morning of 9/11, commander of the same duty station where he began his Coast Guard career thirty years earlier. As a senior officer with years of "M" experience, Naccara had long recognized the need to get the Marine Safety and Operations—the "M" and "O"—sides of the Coast Guard together. For more than a year prior to 9/11, he had proposals in to Coast Guard Headquarters to combine MSO and Group Boston into an Activities command similar to Activities New York at Fort Wadsworth. Another proposal for an Activities Maine would merge two Groups and an MSO in order to provide more efficient coverage of that pivotal Coast Guard state.

When the first aircraft hit the World Trade Center, Naccara had called down to Activities New York to find Admiral Bennis. Told he was heading south on leave that morning, Naccara finally reached Bennis on his cell phone. He told Bennis that he might want to turn around and head back to New York, something Bennis was already doing, blasting up I-95 at nearly one hundred miles an hour. After the second aircraft struck, Naccara called all of his senior staff into his conference room to plan for any possible Coast Guard involvement. When the North Tower came down and the Pentagon was hit within minutes of each other, Naccara felt these events like hammer blows. A native of northern New Jersey, he had been stationed on Governor's Island, New York, in the 1970s as the new World Trade Center was being completed.

Naccara's staff in Boston did not yet know that two of the aircraft had been hijacked out of Boston's Logan Airport. But with the attack on the Pentagon, they realized the pattern of the battle, and started to inventory famous First District sites that might become targets. At the top of this list were the Statue of Liberty in New York Harbor and the Prudential Building in Boston, followed by critical infrastructure, especially nuclear and other power plants located along waterways, as well as important bridges and tunnels. Captain W. "Russ" Webster, Nacarra's Chief of Operations, remembered that within the first hour, force protection conditions (FPCON) were upgraded across the District. Throughout the day, FPCON continued to rise, from peacetime conditions at dawn to full war footing by sunset.

Naccara's immediate problem was to decide not only what he needed to protect, but what the definition of protection was going to be.

> We had so many things to protect, that my question to my staff was, 'How are we going to protect it? What does protect and defend mean?' We were sending our folks out there with a blue light and little else, and I said that I can't accept that. The fact that we're there is somewhat of a deterrent, but in reality it's meaningless. We were pretty much unarmed, and of those who had arms most were untrained. We had no defined rules of engagement. There were just too many questions to be answered, so what do we want to do? We debated that for quite a while.

For Captain Webster, the Coast Guard's responsibility was to maintain its posture as the continuous federal presence along the local waterfront. Regardless of the circumstance, be it 9/11, or an oil spill, or a plane crash, the Coast Guard provided initial disaster response on the water. "The public is used to seeing the Coast Guard first on the scene in a leadership role on the water. And we have a dual role, as the Service that both plucks you from the water when you're in trouble and acts as the cop on the beat, the Smokey on the sea."

*   *   *

For the helicopter crews of Coast Guard Air Station Cape Cod, the first response had been both instinctive and jarring. Trained to pluck drowning mariners from the sea, they were ordered to speed to Manhattan in a dramatic attempt to save people clinging from the burning towers, then ordered to land before they got the chance.

Joseph R. Palfy, a Captain in the Canadian Air Force, was assigned to the Air Station as part of an exchange program that offered Canadian search and rescue helicopter pilots experience flying with the U.S. Coast Guard. Halfway through his four-year Coast Guard tour on the morning of September 11th, he had already flown more than sixty SAR cases with the station's powerful HH-60J Sikorsky Jayhawk helicopters. He remembered the 11th as a beautiful morning on Cape Cod, as he prepared to do a ground "run-up" on one of the Sikorskys to verify that the helo was serviceable and ready to fly. With all the other pilots busy, Palfy called Lieutenant Christopher L. Kluckhuhn at home and asked him to come in and assist with the ground turn. Kluckhuhn arrived a short time later. The engines were started, the rotors spun, and the machine checked for leaks or other potential problems.

A short distance away, Captain Richard P. Yatto, the Station Commander, was in his office when he received a call just before 0900 telling him to go to the Operations Center and turn on CNN. Yatto, a twenty-five-year Coast Guard veteran, looked at the burning North Tower and then at the clear blue skies in the background and knew right away that something had gone terribly wrong. As he watched the impact of the second aircraft into the South Tower, he realized that this was no accident. Another of the station's pilots, Lieutenant Kurt R. Kupersmith, was standing near Yatto, and saw the Captain's whole face change when the second plane went in.

Yatto was immediately analyzing the likelihood of people being trapped on the floors above the levels where the two aircraft had hit the towers. He remembered a hotel fire in San Juan, Puerto

Rico, where people were trapped on the roof and couldn't get off. Those people had died when they couldn't get off the rooftop. "My first thought was that we needed to get a helicopter down there to take people off the roof if in fact people tried to get up on the rooftop."

Yatto asked the Operations Duty Officer (ODO) to call the First District Command Center (CC) in Boston to get their take on the situation. As air station commanding officer, Yatto had the authority to launch his helos, but the usual procedure dictated that his station would be tasked for a specific mission by the CC in Boston. But Boston was in a monitoring mode, and Yatto had the San Juan hotel fire at the forefront of his mind.

"What was their answer?" Yatto asked his ODO. "Do they want us to launch?"

"Well, sir, they're *evaluating*."

"Bull*shit*," snapped Yatto. "We're going to launch right now. Tell them we're going to launch."

Out on the tarmac, Captain Palfy and Lieutentant Kluckhuhn were nearing the end of their ground run when two F-15 Eagle fighter jets from the adjacent 102nd Fighter Wing at Otis Air National Guard base suddenly blasted into the air. The two helicopter pilots noticed right away that something was up. "Usually when they take off," said Palfy, "they turn their afterburners off shortly after the end of the runway. Two of the F-15s took off and they left their afterburners on the entire time." Less than two minutes later, a stream of fighter jets followed hard on the first two, and again none of those fighters turned off their afterburners.

Palfy then heard a call come over his radio. It was Captain Yatto, recalling each of Air Station Cape Cod's helicopters. Two of the helos were conducting standardization ("stan") training flights that morning. But precisely because they were doing "stan" checks, these two helos had light fuel loads, and as a result did not possess the range to reach New York. Yatto ordered these to return immediately. It was then that Palfy spoke up. He and Lieutenant Kluckhuhn were ready to go, they had a flight mechanic, Aviation Maintenance Technician First Class Louis B. Williams, on board, and plenty of fuel. With a full load of 4,000 pounds of fuel, Palfy's H-60 possessed about four hours flying time, and a range of more than 350 miles. At full speed, they could be in New York in an hour.

Neither Palfy nor Kluckhuhn as yet knew why they were being asked to fly to New York. They put on their life vests and waited for a rescue swimmer to be assigned to their flight. Palfy then remembered that Kluckhuhn was trying to qualify as an aircraft commander, so he offered his right-hand pilot's seat to Kluckhuhn. Kluckhuhn would later remember this as one of the most unselfish acts a fellow pilot had ever done for him. As the rescue swimmer, Aviation Survival Technician Third Class Matthew P. Malneritch, ran to the helo and tossed his gear on board, Palfy heard another radio message. It was something about a plane crashing into the World Trade Center. Palfy thought it sounded vague, and thought that maybe a fighter had gone off course somehow. As the Sikorsky lifted off, the crew was now told that two jets had crashed into the Trade Center.

Lieutenant Kluckhuhn drove the Sikorsky to its maximum continuous power, and soon the helicopter was screaming across Rhode Island Sound at 157 knots. A few moments afterwards, Lieutenant Kupersmith followed in a second Sikorsky. The crews began to work through how they might approach the Trade Center, still without fully understanding the nature of the attacks. They talked about their fuel consumption, and where they might get fuel if they had to hover over New York for any length of time. Then they discussed how they might go about plucking people off the rooftops of the towers. They considered their standard rescue load of six people, and compared it with the potential of forty to fifty thousand people in the two towers. Palfy thought that they could attempt to pick people out of the windows of the buildings if they could get close enough. "Worse case scenario, if people were stuck, we could lower our basket down to the upwind side of the building. The winds were out of the north, so we would have approached the north side of the north tower, kept the helicopter in clean air, with the smoke drifting away from us, and lowered the basket."

As the Sikorsky reached Long Island Sound, the crew tuned in the New York Fire and Police frequency, and heard that the South

Tower had just collapsed. They raced on, now just a half hour from the city, determined to reach the still-standing North Tower, assess the situation, and do what they could. They tried to raise New York air traffic controllers but the radio was full of chatter. Approaching Gabreski Field on Long Island, the crew finally raised New York air controllers, who ordered the Sikorsky to land immediately. "We said 'We're the rescue helicopter,'" said Palfy. The helicopter was again ordered to land immediately. Palfy told them that they were a Coast Guard helicopter en route to the Trade Center for a rescue mission. No matter, they were told, all traffic was being closed down, and no traffic was being allowed into New York. The Air Force F-15s, they were told, were clearing the skies and wanted no additional air traffic near New York. Palfy again raised the issue that the Sikorsky was the rescue helicopter. Again they were ordered down.

"To be blunt about it, I was a little pissed," Palfy recalled. With no other options, he reluctantly ordered Kluckhuhn to land the helo. Typically, Kluckhuhn would come in to an airport at 100 to 120 knots, and make a controlled approach to the landing. On 9/11, he came in at 155 knots, brought the nose of the helo up to thirty degrees, and ordered the tower to have a fuel truck standing by.

Palfy told the crew to refuel the helicopter, as he still figured on reaching the North Tower and staying on scene as long as possible. Kupersmith asked his crew to strip their helo, and prepare as much room as possible for potential casualty litters. With Kluckhuhn, Palfy ran into the control building at Gabreski, intent on cursing their way to New York if necessary. For the first time, they saw the burning North Tower on television. Within a few minutes of walking into the control facility at Gabreski, the North Tower collapsed.

The Coast Guard air crew was devastated. Lieutenant Kluckhuhn felt like he had been punched in the stomach. Some remained embittered long after 9/11. They felt that if they had been given the chance to reach the towers they had a better than even chance to rescue at least some of the victims there. "If the buildings were still standing and there were people on the roof, whether or not it was feasible, we would have tried," said Captain Palfy. "We would have put ourselves *in extremis*," said Kluckhuhn, "for a chance to rescue those people." What hurt the most, in the days that followed, was seeing images of people jumping from windows, the very people the Coast Guard aviators would have tried to save.

Their commander, Captain Yatto, agreed. "When I looked again, before the towers collapsed, I think it would have been possible from a certain angle to be slightly upwind. It didn't look like there was a lot of wind that day. There was just enough that there would have been a clear corner of that rooftop." Yatto had full confidence in his crews, knowing that the individual aircraft commander makes a decision once he arrives on scene on whether or not they feel comfortable with a potential rescue. "We do medevacs off fishing boats, and the pilot will look at the boat, look at the rigging, and decide where is the best place to do the hoist to take a fisherman off."

Other issues would have confronted Palfy and Kluckhuhn had they reached New York and been in a position to rescue people off the rooftop or from the windows of the North Tower. A rescue swimmer on a recent Coast Guard air rescue, lowered down to the deck of a sinking former cruise ship, was threatened at knifepoint if he didn't take certain crew members first. Not only did the swimmer calm that situation, but the helo crew, trained to lift as many as six victims at one time, managed to rescue *more than twenty* survivors in one mission. Even with a potential mass of people clamoring to be rescued, Yatto believed that the two Sikorsky's could have made several runs between the towers and safety. "There's a helicopter landing pad—the Wall Street pad—on the East River and that's just a two minute flight from the top of the World Trade Center."

> We do cliff rescue training, we send our flight crews out to Coast Guard Air Station Astoria where they have advanced rescue swimmer training, and they practice things like cliff rescues where you send a rescue swimmer down on a harness and he can pick up a survivor. So it was potentially feasible for our crews to lower their rescue swimmers down fifty, sixty, eighty feet down the side of the building, with the helicopter hovering over the rooftop, to pick someone out of a window.

Palfy, too, thought he would have used the Wall Street pad, but he even considered the possibility of using the deck of the retired aircraft carrier U.S.S. *Intrepid,* now a museum, as a potential landing pad where he could have shuttled survivors picked from the World Trade Center. The power of the 22,000 pound Sikorsky was also something that gave the crew the feeling that they could have succeeded if given the chance. "If anybody was going to be doing any rescuing that day," said Palfy, "it was going to be us."

Flying back to Air Station Cape Cod the following day, Lieutenant Kluckhuhn remembered looking down at the industrial buildings, the suburban homes, and the malls of the Northeast. Everywhere, American flags had suddenly appeared. "Those bastards," he thought. "They have no idea of the strength of this country. They did this one act, and it had a tremendous impact, but it was so minute in comparison to the strength of the whole country."

*   *   *

As the crisis developed at the World Trade Center, Admiral Naccara decided to move more of the district's people and platforms to New York Harbor. When the Towers came down, destroying communications between the district and Activities New York, Naccarra decided that one of those people would be the district commander himself. By the end of 11th, communications consisted of cell phone contact with one or two people at Fort Wadsworth. Getting as many assets as possible on scene, reestablishing communications with Activities New York through Station Sandy Hook, evacuating Manhattan, and reassuring the public that the waterways were secure were the top priorities in the first twelve hours.

Naccara went to New York to establish a Regional Incident Command (RIC) by bringing a communications suite, along with key legal, operational, and marine safety staffers, from Boston. The need to reestablish communications stemmed from the district's need to be able to move assets as required on scene, and as well from the Coast Guard's organizational thirst for information. A Service engaged in continuous daily operations, one that left little time for either contemplation or introduction of alternative strategies for conduct once an operation was underway, the Coast Guard required a continual stream of data passing from district officers to field commanders to field units.

By mid-day on 9/12, Naccara was on board a Coast Guard HH-60J Sikorsky helicopter en route Staten Island and Activities New York. Admiral Bennis was there to meet him and report on the damage that had been sustained. The catastrophic casualty estimates of the day before were coming down dramatically, from 25,000 to 10,000 to 6,000, and the immediate discussions centered on, if and when a U.S. Navy aircraft carrier and battle group might enter the harbor. Neither Naccara nor Bennis thought it necessary. From a force protection standpoint it would have created a nightmare for the Coast Guard, and distracted resources needed at that moment in the rescue and recovery efforts at Ground Zero.

Naccara and Bennis took a small boat across the harbor. The city was eerily quiet, something it had never been during his years growing up nearby. He saw four Coast Guard cutters stationed at different intervals in the harbor. There were cutters on the East River and cutters by the Verrazano Bridge. And that was all. There was no other movement on the water. A couple of F-15s patrolled overhead. Their small boat tied up at the Lower West Side, and from there Naccara and Bennis walked to Ground Zero. Naccara was in disbelief.

> The people that were walking around were dazed. I walked to where people were cycling in and out of the hole, and people were speechless. The workers were covered with grit, from the powdered cement which covered everything, three-quarters of an inch deep on the cars where we walked. Yet it was amazing what endured, along with what was destroyed. Everywhere you stepped there was paper, all kinds of personal notes, just laying in the street. Emergency vehicles were destroyed, parts laying about. It was the most devastating scene I'd ever seen, and hope to ever see, in my life. Yet no one wanted to leave or take a break.

No one was wearing any kind of respiratory equipment, and the thin paper masks worn by Naccara's party were ineffectual against the stench in the air. Eventually, the masks came off and were tossed aside. Naccara walked to a Coast Guard buoy tender that was providing water and food to New York firefighters who were cycling in and out of Ground Zero. The normal reluctance of Coast Guard personnel to say what was on their minds whenever an admiral was around had vanished. Naccara made a point of his New York roots early in every conversation, breaking the ice by letting people know that the attacks affected him in a personal way, too. It seemed that everyone he talked to was extremely distraught. So he conceived of his role as one of maintaining an appearance of calm in a storm.

> It's one of those things that you learn, in your education as a flag officer, and it doesn't happen overnight. And sometimes the magnitude of that expectation might surprise you, if that's not your typical makeup. People look to an admiral, at any level, for certain levels of reassurance, especially in a crisis. I felt it was important that I be seen in control of myself and my senses. So we would talk about issues such as 'what can the Coast Guard do to help? How can I bring more assets into New York to help?' And I needed to explain to people that it was not good to be there for thirty-six hours straight. You need to take a break. You need to take a rest. But everyone wanted to help. No one wanted to go home.

If Naccara himself had a need at this moment, it was for hard intelligence concerning potential threats to other areas of the First District. Such intelligence, from any source, was extremely sketchy and very hard to come by. Naccara went through local FBI contacts, through the intel staffs at Atlantic Area and Headquarters, anyplace he could think of to try to gain ballast for his operational plans and resource allocations. In the absence of such data, it was quickly apparent that Coast Guard Reserve Port Security Units (PSUs) would be needed on scene as fast as possible.

In his role as Regional Incident Commander, Naccara was given two PSUs (305 and 307) for the entire First District. Even as they were en route north, he dispersed these highly-mobile, highly-trained and heavily-armed units, 305 to New York and 307 to Boston, to cover as much of the District as possible. With the continuous nature of the operation on the ground, and facing a dearth of good intelligence, the constant requests for information from up the chain of command rapidly became a distraction.

> I tried to depict the scene as well as I could, including personal observations, discussions with other key folks, and tried to describe the ineffectual communications we had in place. It took a special cell phone just to communicate with the Commandant or the Area Commander, we couldn't communicate by our normal methods. If I was looking for anything from them, I wanted them to tell me that if I needed additional assets, they would provide them to me. When I look to my Area Commander, I would like him to respect my requests, for example, if I wanted a couple of 270s to come in off drug interdiction or living marine resource work, I think he should listen to what I have to say and my explanations and then respond. And that's the way [Admiral Allen] worked. He was very responsive to our needs.

To identify critical infrastructure across the district, Naccara's staff devised a three-tiered system of threat ranking, and resources would then, in this construct, flow to the highest-tiered facilities. In practice, however, the impulse of the field commanders was to come back to the district with a list that placed every asset within a particular Group or MSO in the first tier. This was clearly unworkable, so Naccara's staff set about to prioritize infrastructure within the district, leading to some heated discussions. But the issue in the end was moot, because the Coast Guard possessed nothing approaching the level of operating assets—cutters, aircraft, and small boats—with which to protect everything.

Based on this analysis, and armed with a chart that showed the Coast Guard units and personnel available throughout the district, Naccara set out to meet with the governors of each state in his district. He offered this data to each of the governors, explaining the level of forces they could expect in peacetime, as well as that in place after 9/11. He showed which tiered assets the Coast Guard had chosen to safeguard, and asked for assistance from the National Guard or other state and local forces to look after the others. These would become instant port security force-multipliers for the Coast Guard, which would never have the personnel nor in fact the mission to provide physical security for each and every port in the United States.

These new levels of intergovernmental cooperation after 9/11 became especially helpful in late September and early October when a series of liquefied natural gas (LNG) deliveries into Boston Harbor were delayed by concerns over security and public safety. The Coast Guard had ordered an LNG tanker out of Boston Harbor on 9/11 for these same reasons. The issue preoccupied much of Naccara's time, having to explain to the public and press alike that, as fearsome as they might appear as a terrorist weapon, an LNG tanker was in fact largely the opposite. Even if you were to some-how blow a hole in an LNG tank, apart from any casualties from a localized fire in the immediate area of the explosion, the resulting gas leak would likely do little more than dissipate into the atmos-phere, with little or no danger to the general population.

Dr. Alan S. Schneider, a civilian analyst for the Coast Guard who had been involved with the LNG issue on and off for almost thir-ty years, studied it again after 9/11. For Schneider, the amount of energy released in the event an LNG tank was breached would not be enough to devastate large areas. LNG cannot detonate in an unconfined space, unlike other hazardous chemicals carried by tankers, and poses few of the mass poisoning dangers inherent in the release of other gases like chlorine or ammonia. The terror dan-gers from an LNG tanker were on an order of magnitude less, in Schneider's thinking, than atomic weapons or even a jetliner flying into the World Trade Center.

But the issue of LNG tanker safety had been a part of the Boston waterfront security plans and, by extension, Boston politics, for thirty years. So despite Naccara's efforts at public and political education regarding the tankers as an unlikely terror weapon, the Coast Guard was forced to pay far more attention to LNG tanker movements than Naccara thought prudent. Establishing enhanced security zones around the tankers played, in effect, into the hands of the terrorists, by multiplying the fears of the public while at the same time distracting extremely limited Coast Guard resources from other, higher tier, threats to the waterfront.

On the other hand, such extremely high visibility events went a long way toward the goal of reassuring the public that was such an integral part of Coast Guard operations after 9/11. Ten days after the attacks, Naccara was getting out of a government car and heading to his office when he was approached by a stranger on the street outside First District headquarters.

> He ran over to me and said 'Thank you so much. I can't thank the Coast Guard enough for what you have done for the city of Boston. Just having your presence out on the water makes us all feel so much safer.' And then he just walked away quickly. Riding the train to work, where people know from my bags that I work for the Coast Guard, people have said to me that they didn't care so much about the LNG issue, so much as they just wanted to see the Coast Guard on the water.

Coast Guard personnel were offered free meals or hair cuts. Port Security Unit 307 personnel were treated to an evening of food and entertainment by a collection of restaurants from Boston's North End. Naccara saw this as an affirmation of the Service's role in calming public fears in a crisis. This was a turnabout for many Coast Guard personnel, especially those who had spent years on cutters on patrol at sea, bearing the brunt of abuse from mariners who traditionally rejected any form of rule or regulation.

Perhaps most remarkable was a suggestion by the Governor of Maine that the Coast Guard use the fishing fleet as, in effect, a first line of defense offshore, an idea not seen in the First District since the Second World War. Naccara was once again quick to multiply his force. A "Coast Watch" program was created, with an 800 number for anyone to report any anomaly they perceived along the waterfront or in the near coast area. A "Coast Picket" program, using fishing vessels to call in any suspicious offshore activities to either the FBI or Coast Guard, expanded from Maine down through Massachusetts and Rhode Island.

The fact that terrorists had been casing the Boston area for weeks, if not months, prior to hijacking two aircraft out of Logan Airport, led Naccara and his staff to consider several of these and other new ideas for gathering intelligence along the coast. Which vessels to board, which passenger lists to scrutinize, whom to detain and on what grounds, especially in a harbor area filled with high speed ferries and a constant movement of people, all such programs were studied and aligned with what limited manpower the district had at its disposal.

Without the resources to blanket the waterfront, the problem came down to better intelligence of potential threats against high-tier facilities. So for Naccara, the issue for the Coast Guard came down to investing in more and better intelligence, shared more efficiently and effectively both inside the Service and with other government agencies. After 9/11, he learned that the Navy flew twice daily P-3 sorties from Maine to Virginia. Could those flights be used to report in real time on vessels of interest to the Coast Guard? Naccara thought that policies could be developed that would reconcile such data transfers with the federal doctrine of posse comitatus, which restricted the use of federal armed force in the pursuit and capture of civil violators of law.

In the movement of container ships, alone, the First District initiated several new programs designed to enhance both container security and border security. Operation Safe Commerce focused on the movement of a single container vessel, from Europe to Canada and across the northern border to the ports of the northeast, to study the movement of a container through its multi-modal voyage, through varying levels of security. A New York-New Jersey Megaport Project enlisted more than a dozen shipping companies in an effort to identify better methods of waterfront security.

Such studies sought in large measure to generate strategies for preventing the introduction of a weapon of mass destruction through the American waterfront. Naccara believed that should such a weapon find its way to an American port, then it was already too late to do much about it. All the sophisticated monitoring equipment in the world, all of the Coast Guard's famous crisis response capabilities, would not prevent the devastation that would follow. The alternative was to go back to the point of origin, to track ships and cargo from their sources. As for the Coast Guard itself, Naccara believed that 9/11 showed that the Service must form the maritime core of any new border or homeland security agency. To do otherwise would be to leave the Service behind, as a kind of lesser coastal emergency reaction force, rather than a proactive frontline border defense organization.

> In the weeks following 9/11, we tried to reclassify all our missions, counter-drug, fisheries, and so forth, as a part of homeland security, with port security being only one slice of our operations. It would be damaging to the Coast Guard if you were to break us up in any way. The whole Coast Guard should remain an entity, since each of our missions enhances homeland security.

When the President visited Portland, Maine, early in 2002 to thank the Coast Guard, he was taken into the Combat Information Center on board the cutter *Tahoma*, where he saw a couple of radar repeaters showing an outline of the New England coastline. He asked if this enabled the Coast Guard to see all the ships offshore and the answer was no, the Service doesn't have that kind of a system available. But, said Naccara, there was a potential fix, and it involved fishing vessels who already participated in the Voluntary Management System (VMS). In the First District, some 300 fishing vessels voluntarily carried a transponder which allowed the National Marine Fisheries Service and the Coast Guard to view these vessels in real time. The Coast Guard could watch as the vessels cruised along the edges of closed fisheries areas, perhaps cutting a corner or two here and there. It allowed the Service another tool to increase its awareness of the maritime domain.

When the President heard this, he asked why the U.S. couldn't have one of these transponders on every ship? It was a chance for Naccara to say that the Coast Guard had been trying to do this for years, with the Service inserting such capability into its new Vessel Traffic Service centers coming on line along the Mississippi River and all other U.S. pilotage waters. But on the maritime frontier, such notions were still far off.

But there was no question that 9/11 had changed the whole equation. Previously wary fishermen, Naccara was quick to observe, now wanted the Coast Guard around. The Coast Guard had always been there as a search and rescue presence. Now they were there, Coast Guard personnel and fishermen both, as guardians of the maritime frontier, partners in a common cause.

Chapter Four

9/11 and the Area Commanders:

Part One: Thad Allen (LANTAREA) as Resource Broker

Vice Admiral Thad W. Allen was at the doctor's office. It was one of the few open times his staff had been able to find in a September schedule that included over one hundred and fifty meetings, briefings, teleconferences, commissionings, decommissionings, retirements, welcome aboards, courtesy visits, award ceremonies, and a host of other administrative necessities. Other than the hour for his physical at 0700 on 9/11, his calendar had an hour reserved for "spouse" on a Saturday afternoon later in what was shaping up as a normal month for the Commander of the Coast Guard's Atlantic Area, its Fifth Coast Guard District, and its Maritime Defense Zone Atlantic.

Thad Allen had all the gentle persona of a hurricane inside a volcano. As he himself admitted, he was something of an anomaly in the flag corps, Popeye in a Service that tended to favor tall, slim officers. The Admiral looked more like the classic notion of a Chief Petty Officer, which his father, Coast Guard Chief Damage Controlman Clyde Allen was during the Second World War. Graduating from the Coast Guard Academy in 1971, Thad Allen commanded a LORAN Station in Lampang, Thailand, during the last years of the Vietnam War. As a Commander, he spent a year at the Alfred P. Sloan School of Management at the Massachusetts Institute of Technology, authoring a 350-page thesis entitled *The Evolution of Federal Drug Enforcement and the United States Coast Guard's Interdiction Mission.* Such historical perspective was put to use in Allen's immediate past assignment, as Commander of the Seventh Coast Guard District in Miami, where his forces had seized fifty tons of cocaine.

Allen assumed command of the Coast Guard Atlantic Area, an area of operations spanning over fourteen million square miles, on June 27, 2001. At the time, migrant and counter-drug operations—mostly run out of Allen's old command in Miami—were the two main stress factors on Atlantic Area's operational platforms. The kinds of port security operations that kicked in after 9/11 were barely on the radar screen. "Any preplanning we might have done for one of these types of events before 9/11, we didn't execute to those plans on 9/11."

When the second aircraft crashed into the World Trade Center, Allen hurried back to LANTAREA's Command Center in an attempt to gain, as the Coast Guard liked to call it, 'situational awareness.' He created an Incident Management Team, with different cells devoted to tracking different areas of interest, and within twenty-four hours had stood up an ad hoc homeland security cell. Each district within his Area did the same. The Coast Guard command structure gave wide latitude to its district commanders for events within their operational areas. The Captains of the Port likewise possessed high levels of authority to close ports and waterways, secure important facilities, and enforce a wide array of maritime laws and regulations.

The Captains of the Port were already in motion. Rear Admiral Richard Bennis and Activities New York had closed New York Harbor, while Captain Roger B. Peoples in Baltimore had closed that harbor as well as the Potomac River above the Woodrow Wilson Bridge. In classic Coast Guard fashion, the Service was surging all its assets towards an emergency, until it could gain situational awareness in order to stay on mission for the long haul or to start to peel back some of the resources on scene or en route.

Amongst its other manifestations, the Oil Pollution Act of 1990 (OPA-90) had stood up a National Incident Command System (NICS). The NICS mandated that the Coast Guard send a senior flag officer to oversee and manage any incident of "national significance." By mid-morning, Allen had had several telephone conversations with Admiral Naccara in Boston, in part to discuss whether or not Naccara should move from Boston to New York. Ultimately, Naccara did move to New York, but just for a few days, before migrating back to Boston. The Coast Guard decided to leave command and control at the local, Activities New York level, primarily because the Coast Guard already had Admiral Bennis on scene, a personnel anomaly brought about in part by Bennis' ongoing battle with cancer.

The need for Service-wide situational awareness quickly led to a new system of communication between the senior leadership of the Coast Guard. Admiral Allen appreciated the immediate and intensive levels of communication:

> What evolved within a matter of hours, we shifted to a twice-a-day conference call with all the impacted parties. After a few days it became a single daily call, at around four in the afternoon. What I did was to have a call at two o'clock with all my district commanders. We'd start out with Admiral Bennis, then Admiral Naccara giving an update, then everybody else would report on what was going on. When that was done we'd roll into a larger teleconference with the Commandant, along with Admirals Bennis and Naccara. For the first week to ten days that was a daily occurrence, so we were really plugged in tight. We were doing things with total visibility up and down the organization.

Allen faced several conflicting impulses that morning. There was the impact to the national psyche to be considered, then the organizational response of the Coast Guard. His first impulse was to fall back on his own experiences dealing with major oil spills as a Captain of the Port of Long Island Sound. The maxim he had followed in those cases, when oil was hitting the water and it was clear it was going to be a significant event, was to take half his people and send them home. Otherwise, twenty-four to thirty-six hours later, the whole staff would drop from exhaustion. Within an hour of the attacks, Allen became concerned about the Service's ability to sustain a major operation for any extended period of time, with the need to close and secure New York Harbor and what was going to be involved thereafter.

In a further attempt to gain situational awareness, Allen diverted every major cutter underway on 9/11 to a port. He then got some cutters underway that had not been on patrol. Three large cutters, including the two Bear Class medium endurance cutters from New Bedford, Massachusetts, *Campbell* (WMEC 909) and *Tahoma* (WMEC 908), and the Juniper Class seagoing buoy tender *Juniper* (WLB 201), from Newport, Rhode Island, ended up in New York Harbor. They would be joined eventually by another thirteen cutters and more than thirty small boats.

Allen needed these platforms in ports up and down the East Coast for more than intelligence and security. These operational platforms possessed command and control architecture—secure communications and heavy weaponry—far beyond anything a Captain of the Port might have access to. In the Coast Guard command structure, the COTPs had legal authority over U.S. ports and waterways, but no operational platforms with which to enforce that authority on the water unless the COTP happened by geography or convenience to be co-located with the Coast Guard group or district offices that did retain control over operational assets. Many Captains of the Port operate from leased space in an office building, and communications often amounts to little more than a Nextel phone and a VHF-FM radio. This reflects the difference between the two major operational programs in the Service, 'M' and 'O,' which has existed since the early 1970s.

One prominent command where this separation had been merged was at Admiral Bennis' command, Activities New York. But Activities New York's communications had been destroyed on 9/11. The cutters brought Bennis an instant communications suite, acted as an extension of his eyes and ears as Captain of the Port, and executed tasking as he required. Allen's first bridging strategy on 9/11 was to get those cutters in place within the ports, so that the Coast Guard would have command and control capabilities up and down the East Coast.

9/11 exposed a few of these corporate Achilles' heels. We had done some experiments in New York and elsewhere establishing combined "O" and "M" commands. I told my staff that if this had to happen, it was good that it happened in New York, because there we had the right command for the response that we needed in the harbor that day. But we have other places—and the poster children are Wilmington and Savannah—where you have Captains of the Port of major ports, which include load-out ports for military operations, that aren't co-located with a Group, and the nearest operational unit is a small boat station twelve miles down at the mouth of the river. It was virtually impossible for those places to carry out any kind of a coordinated operation because they did not have either the capability or command and control. Therefore, cutters were the immediate solution.

By mid-afternoon on 9/11, Allen came to believe that the attacks were most likely an anomalous situation, a unique event. But he couldn't discount the possibility that they were part of an even larger coordinated attack that might involve the maritime sector. So he reacted tactically, on the assumption that there might be more such events coming somewhere on the waterfront. The grounding of the air fleet removed the potential concerns that might have come from that arena. And the second and third order consequences of the 9/11 attacks—the evacuation of Manhattan, the proximity of the World Trade Center and the Pentagon to the water—these placed 9/11 into the maritime, and hence Coast Guard, domain.

As a field commander, he understood that as the dangers or risks of an operation began to mount, it was a commander's imperative to retain his ability for clear-headed, almost clinical detachment. Such detachment was not shared universally. The daily phone conferences became a kind of anthropological laboratory, as field commanders put forth their best estimates of threatened facilities within their own areas of responsibility, and argued for more resources in order to protect them.

The Coast Guard's response to 9/11 was port-centric, because the Captain of the Port authority was the central authority we were dealing with in our response. Everybody immediately became aware, intuitively and by a visual inspection of their vulnerabilities, of the risks involved. Savannah has an LNG facility, Hampton Roads a major U.S. Navy load-out port. If you're in Miami, you have five million cruise ship passengers a year. If you are in the Eighth District, you travel down the Houston Ship Channel past twenty-three miles of potential incendiary devices. On the Great Lakes it was the Sears Tower and the vulnerability of Chicago on the waterfront. Everybody took what they saw, created a mental template, then applied it to their own port and said, alright, what would happen if it happened here. And that was driving a lot of behavior.

This went to the heart of Allen's job as Area Commander: the need to resolve resource requests across districts, and to funnel the resources that he himself controlled—long-range aircraft and cutters—to where they were needed most. If there was frustration in the districts, some of it could be traced to the Coast Guard's inability to "militarize" its operations overnight. Within days of 9/11, Admiral Allen began to preach to those around him that "water is different." With 95,000 miles of coastline, the Coast Guard was forced to defend a ubiquitous frontier. If you discounted accidents, or the fact that a fanatic can fly a plane into a building, or that an out-of-gas Cessna might land on a freeway, then planes only take off and land at airports. Railcars stay on rails, and cars for the most part stay on roads. As Allen reminded many audiences, it takes only an inner tube and a pair of flippers to enter the Coast Guard's world of work. A porous maritime border, further opened by a long history of freedom of marine navigation, meant that water was a qualitatively different medium to guard than air or land. A cargo ship full of terrorists could transit from Halifax, Nova Scotia, to Havana, Cuba, and as long as it remained fifteen miles offshore and gave the Coast Guard no reason to board, that ship could make that transit. These were national maritime policy issues the Coast Guard would never solve, but instead had to work around. And they were issues that 9/11 sprung wide open.

Immediately, situation reports were generated by the Incident Management Teams or Crisis Action Cells in each of Allen's districts—the First (Boston), Fifth (Portsmouth), Seventh (Miami), Eighth (New Orleans), and Ninth (Cleveland). A command and control apparatus developed within twelve hours.   At 0608 on September 12th, Allen was able to send a message to his Fifth

District commands, offering a district-wide status report, a summary of the actions that were being taken, and plans and recommendations for the next twelve hours of operations.

From his other district commanders, he required a command and control scheme that would allow the Captains of the Port to execute their tasks. If Allen was going to assign resources to them, he first wanted to know how the district commanders planned to manage those resources and exert command and control over them. Once he had that template from each district, he would know where to assign resources; if no template arrived on his desk, that district had a slim chance of prying platforms out of the Area command. Allen did not mandate the structure of these templates, in part because there existed no uniform Coast Guard command structure upon which to lay such a construct.

> I had Activities, I had Groups and Marine Safety Offices (MSOs) co-located, and then I had them where they weren't co-located. But I said, 'You give me a task execution model, and then I will tell you how we're going to distribute the resources.' And they all came back with a slightly different model that basically did the same thing. For instance, if you went to the Great Lakes in the Ninth District, from the unique geographical boundaries there they created "sectors," and within those sectors Admiral Hull designated one person to be sector commander. That became their organizational structure: they fused and unified their commands, and all the operational reporting came out of that sector command. In some cases it was an MSO C.O., and in some a Group Commander. It depended on where they were located and their capability. In the Eighth District, Admiral Casto created Crisis Task Units (CTU). So if you were in Houston-Galveston, he took MSO Port Arthur, MSO Houston, Group Galveston, and VTS Houston, put them together and had them operate as a task unit. And each of these units had some number behind it. So if I wanted to send a cutter down, to station it off Houston-Galveston, I could chop it to CTU xxx.x.

In Allen's own Fifth District, he divided his operational response by Captain of the Port zones. When the Captains of the Port came back to him with a plan to execute their security mission, Allen would tell them what resources they could expect from Atlantic Area. His first priority was to speed cutters to those ports without

Area. His first priority was to speed cutters to those ports without adequate command and control over their operations. The cutters moved out to support the Captains of the Port, to provide them not just with command and control, but with eyes and ears, and with a platform with which to board suspect vessels. The cutters became, in effect, floating operations centers for the Captains of the Port. To Allen, 9/11 and the arrival of cutters within the ports rendered the Coast Guard's entire shore-side command and control structure obsolete.

As well, the ad hoc nature of the various district operational response models exposed the lack of a paradigm for the defense of American ports. The U.S. Navy had always been an expeditionary force that trained to fight "over there." After the wreck of the oil tanker *Exxon Valdez* and the subsequent Oil Pollution Act of 1990 (OPA-90), the Coast Guard had increasingly focused on marine safety and environmental protection, along with its long-standing counter-drug, migrant, and fisheries work. Somewhere in there port security and defense readiness had been allowed to fade somewhat into the background. Command structures like the Maritime Defense Zones, created during the Cold War, were shown to be largely irrelevant to a global war on terrorism.

In Allen's view, 9/11 forced the Service, from the Captains of the Port on up to the Commandant, to begin working through all the issues surrounding the security of the ports that would now be required in its aftermath. Policy groups might spin out theoretical constructs, but now the Coast Guard was collecting a massive data base of real-life experience on all of the complexities involved in guarding the nation's strategic ports and waterways. The senior leadership of the Coast Guard had not dealt with such a multitude of port security problems since the Second World War.

In the early weeks of the crisis, it appeared that much of this operational data would be subsumed by whatever strategic doctrine was eventually adopted by Northern Command, and whatever responsibilities for port security were absorbed by the new Director for Homeland Security. But with the proposal to create a new Department of Homeland Security, with the Coast Guard as a key component, these lessons learned would only increase in value. Such data was put to immediate use in the hastily-accelerated port vulnerability assessments. When it came to defining just what a secure port would require, Allen instructed his staff to come up with a base set of criteria on what constituted port vulnerability. This was then applied to each of the Atlantic Area's ports, in order to produce a port-specific security plan that offered equivalent levels of security, based on each port's unique vulnerabilities and threats.

Allen also leaned on his district and operational commanders to be realistic in assessing their force levels. Based on their unique career paths and experiences, different field commanders reacted to 9/11 in unique ways. Some reacted by throwing their entire force into around-the-clock security patrols, while others preferred to wait and see what threats might arise based on hard intelligence before committing their operational resources. There were no simple answers, and in the absence of hard intelligence of a credible maritime threat, local commanders fell back on personal experiences, visceral reactions to the attacks, or familiar routines that had worked in the past.

The Coast Guard would continue to struggle in such situations with the dual personality of an armed force with civilian law enforcement responsibilities until a new doctrine for behavior was developed. One way to deal with this identity crisis, Allen believed, was to move toward a Marine Corps model where traditional doctrine holds that every Marine is a rifleman. In Coast Guard terms, this would mean that all vessel boardings would be conducted uniformly, with no separate boardings for law enforcement, or marine safety, or recreational boating. There would be one uniform U.S. Coast Guard boarding.

> You have to apply your resources to the highest known threats, vulnerabilities, and consequences. The Coast Guard has done that very thing for many years. If you have a low-probability, high-consequence event [like 9/11], you're going to have to articulate [after the fact] that you were employing your resources towards the highest-probability, highest-consequence event. If you can't do that, if you were winging it, you're going to get hoisted on your own petard.

After 9/11, Allen found no fault with either approach, active or passive. Those who reacted aggressively were quick to point out that there were no credible threats to the World Trade Center from a commercial aircraft before 9/11. And yet it happened. For the Coast Guard, that meant that more than likely there would be no warning of an impending maritime threat. Such thinking placed an enormous, and Admiral Allen would argue, impossible, burden on the Service. Ninety-five thousand miles of shoreline, and a determined enemy who needed nothing more than an inner tube and flippers to enter the Coast Guard's world of work. That was the Coast Guard's new reality.

Chapter Four

9/11 and the Area Commanders:

Part Two: Ray Riutta (PACAREA) and the Centrality of Intelligence to Warfighting

In a minor historical coincidence, both of the Coast Guard's Area Commanders on 9/11, Thad Allen in Portsmouth, Virginia, and Ray Riutta in Alameda, California, had commanded LORAN stations during the Vietnam War; Allen at Lampang, Thailand, and Riutta on a spot of land surrounded by water and rice paddies at the mouth of the Perfume River near the village of Tan My, just forty-miles from the Demilitarized Zone (DMZ).

Riutta's station helped provide navigation guidance for squadrons of B-52 bombers. When those squadrons let go their bomb loads to the north, the resulting ground shock would trip the instrument timers at the Tan My station. Riutta and his men had forty-five seconds to get them back on line or the second and third wave of bombers would lose their navigational locks and be forced to abort their missions. At one point in the spring of 1972, after 15,000 North Vietnam troops attacked and broke the South Vietnamese Division buffering Tan My from the DMZ, Riutta's station was caught in a pocket, with his corpsman tending to battlefield wounded until the area was retaken.

Nearly thirty years later, after a career as a surface operator that culminated with the job as Admiral Loy's Chief of Operations at Headquarters, Riutta was named Commander of the Pacific Area. It was where he wanted to be, having grown up on the Oregon coast, and where he had spent most of his time in the Service, from weather ship duty out of Hawaii to buoy tender work in Alaska, from a patrol boat in Monterey to command of a 378' Hamilton Class high endurance cutter in San Pedro.

On the morning of 9/11, Riutta received a call from his Command Center at about 0600, asking him to turn on his television, which he did in time to see the second plane hit the World Trade Center. The magnitude of the attack surprised the Admiral, but not the attack itself. PACAREA had scheduled a war-game exercise in June of 2001 at the Naval Postgraduate School in Monterey, one that centered on homeland security scenarios on the West Coast, in particular an attack upon the homeland itself. A lack of sponsorship partners, typical of the pre-9/11 atmosphere, had postponed the exercise until November.

Riutta gathered his Chief of Staff and his Chief of Operations. With phone calls now flying between districts, they stood up the Area's Crisis Action Center to begin sorting out what actions would be taken throughout the day. The tightly-knit familiarity of a small naval Service simplified the connections between the Riutta and his district commanders, Rear Admiral Erroll M. Brown in Thirteenth District in Seattle, Rear Admiral Ralph D. Utley in Fourteenth District in Honolulu, and Rear Admiral Thomas J. Barrett in Seventeenth District in Juneau. Riutta had known these commanders throughout his Service career, and his confidence in their abilities enabled him to act as a resource manager for the Area secure in the knowledge that they knew what steps to take in securing their ports.

The first action taken in the Area itself was to close all the major ports. But even as this order was carried out, Riutta knew that the ports could not remain closed for very long. All large cutters were recalled from sea and stationed outside a major port, to begin the process of boarding all incoming vessels. PACAREA had an advantage in that it possessed well-defined choke points. Large vessels had only a few major Pacific ports in which to call, and this shrank the waters the Coast Guard would have to cover. If terrorists could hijack an aircraft and turn it into a weapon of mass destruction, then the Coast Guard had to deflect the equivalent threat to the ports: the use of a large commercial or passenger vessel for mass destruction.

If there was a gaping hole in what the Area could provide, it was hard intelligence. Every district wanted it, and the PACAREA Intelligence Team (PIT), the go-to center for situational awareness, could offer little of it. In the absence of hard threat data coming in, PACAREA began to triage those places where strategic interests dictated they deploy their limited resources. In the first hours of the crisis, this thinking led Riutta and his staff to run down the list of things they wanted to prevent, beginning with massive loss of life and massive disruption to the economy. This had the effect of limiting their scenarios to those involving weapons of mass destruction (WMD), with most other scenarios off the table until they got the big picture sorted out.

This "big picture" revolved around ships carrying hazardous cargo into Pacific ports, critical infrastructure, and cruise ships. Gargantuan cruise ships carrying anywhere from two to four thousand passengers worked the Pacific from Honolulu to Alaska. For a captain of the port, it was the equivalent of a small city landing at his docks. For Riutta and the Coast Guard, such a vessel was very difficult to protect and now became a potential World Trade Center. On September 11th, Riutta at least had one advantage in this regard. The high cruise ship season was over, with the Alaska season largely over, a few ships plying the waters between Mexico and the ports of southern California, and a few ships operating in Hawaii.

The late cruise season narrowed PACAREA's focus to hazardous cargo and critical infrastructure. Riutta's team undertook an analysis, along with the California highway department, to study the effect of a large vessel collision with one of the state's major bridges. The bridges had been built in the 1930s to withstand the impact of the largest ship of the day, but ships had grown in both size and mass.

The hazardous cargo scene was not much better. The Coast Guard's mission after 9/11, as Riutta saw it, was to secure the ports while returning the movement of commerce as quickly as possible. Boarding each and every vessel entering each major port could not be sustained over a long period of time. Lengthy boardings simply held up commerce too long. A dirty bomb or a nuclear weapon could be on a container ship, but the Coast Guard had no way to detect such a weapon, nor the resources to search more than 40,000 containers entering U.S. ports each day. All air traffic had been shut down, but Riutta had no intention of letting that happen in his Pacific ports.

But most of PACAREA's assets were in the deep draft category: high endurance cutters and icebreakers. Most of Riutta's Law Enforcement Detachments (LEDETs) were already deployed to other missions and therefore unavailable. With little ability to pull small boats into harbor security work—though this was also done—the Area's harbor protection assets were minimal. Riutta improvised in two ways, both involving his Reserves and Auxiliary, without which Riutta believed he "would have been dead in the water on September 12th."

The first was to bring both of the Area's Reserve Port Security Units (PSUs) on line immediately; PSUs 311 and 313 were deployed to Long Beach and Seattle, respectively. Riutta did this with a single phone call to Admiral Loy in Washington. Riutta took this action even before formal authorization was given from the Secretary of Transportation. "I called the Commandant and said 'Boss, I've got to do this,' and he said 'Do it, and I'll get the permission from the Secretary.'" It was the kind of interplay made possible when you had known someone for more than thirty years, had gone to war with them before and knew how they thought.

In San Francisco, where Riutta did not have the luxury of a PSU, he took the boarding teams from 378s that were in port in Charlie status, and sent them to augment the local Group Command. A Tactical Law Enforcement Team (TACLET) was brought up from San Diego as well. But his primary strategy was the stand-up of the Sea Marshall program, born very early in the crisis, in response to a simple formula: what can we protect with what we've got.

Sea Marshals were, in the main, Coast Guard Reserves with law enforcement training, who acted to prevent a terrorist from doing with a ship what they had done with an aircraft. They would board a vessel at the sea buoy and, from the sea buoy to port, keep watch over the critical maneuvering stations of the vessel: the pilot-house, the engine room, or the steering gear, depending on the type of vessel. They would use the best intelligence available to target those vessels that might require a full-blown search at sea.

The goal then, was to prevent someone from taking control of a large vessel at the last minute and using it as a weapon of mass destruction. Sea Marshals were not formed, or trained, to retake a vessel already in the hands of a group of determined terrorists. That was U.S. Navy Sea Air Land (SEAL) Team work, and it would have to be done when a vessel was far at sea. The Coast Guard had no real capability to stop a large vessel once inside a U.S. port, and even offshore, a high endurance cutter had little capability to sink a cargo ship without support from the U.S. Navy.

After 9/11, many Americans, remembering that the Coast Guard indeed possessed those capabilities during the Second World War, expected that the Service still retained them. Some of that capability still existed within the TACLETs, in the Coast Guard's counter-narcotics cadres, and in the PSUs with their operations in the Persian Gulf, but the rest had to be rebuilt in a hurry. For Admiral Riutta, that left one path in between the two extremes of fighting terrorists at sea or in the ports: intelligence gathering in support of targeted boardings. This became the main strategy available to the Coast Guard in securing the ports in the Pacific Area.

> We learned a very important lesson on 9/11, in that we had basically been demobilizing our entire port security structure over the past ten years. There was virtually nothing left. None of the MSOs had any capability of talking securely much less handling classified materials. The first few days after 9/11 we had communications folks driving around with STU IIIs [Secure Telecommunications Units] in their cars, handing them out to MSOs just so we could talk secure to them. We had been in the process of taking the weapons away from all the MSOs. None of them were qualified even to use a gun, which presented some problems when we needed to do armed boardings. Guess what, nobody could shoot straight, over at the MSO. It wasn't because they didn't want to, it was because all their weapons were gone and the qualifications had all lapsed.

Using Reserves, the weapons range at Petaluma was hurriedly stood up in an attempt to rapidly recapture lost weapons qualifications. Riutta also brought on board a Reserve Port Security Unit commander to act as a "Red Cell." Whenever Riutta's staff would come up with an idea for increasing security, it was the Red Cell's job to think like a terrorist and tell them how he would defeat the new measures. This had the effect of further focusing the Area's counter-terrorism plans.

Coming from the "Operations" side of the Service, Riutta had been involved in many conversations that lamented the loss of weapons qualifications throughout the Service, and in particular, in the ports. The effort to "streamline" the Service in the 1990s had the effect of forcing commanders not simply to evaluate their local threats, but to take off the table those assets necessary to engage threats that had been downgraded. This led to the loss, not just of weapons qualifications, but of port security and crisis planning in general.

For Riutta, as with so many other senior leaders in the Service, 9/11 revealed the ridiculous nature of the split within the Coast Guard between Operations and Marine Safety. Some of the "M" and "O" dichotomy in PACAREA had been bridged through the creations of Activities San Diego, and the equivalent of Activities commands in Los Angeles/Long Beach, in San Francisco, and in Seattle. Valdez was strictly an "M" outfit, with no surface assets. "I doubt that there are any senior leaders left in the Service who do not believe that the Activities concept is the way we will do business from here on," said Riutta. 9/11 also showed the end result of years of developing partnerships in each of the Area's major ports.

> In L.A./Long Beach, for example, there was no question in anybody's mind who was the acknowledged leader in that port, and they all turned immediately to our Group/MSO captain. The same was true in San Francisco/Oakland. Captains [Larry] Hereth and Captain [Timothy S.] Sullivan became the points around which everyone rallied. In the case of L.A./Long Beach, we had such a good relationship with the port police, with the harbor masters, with the mayors of both cities that they actually

offered police to work right alongside our folks as Sea Marshals to escort vessels in, and provided divers to sweep for mines underneath cruise ships. The partnership in that port is the standard, I believe, for the way we ought to do things everywhere in the country.

Another strategy Riutta employed was to bring on board his Reserve flag officer, Rear Admiral Mary O'Donnell, and make her the Commander of the Eleventh District. This enabled Riutta to concentrate on pressing questions of Area-wide resource management. It also demonstrated to him that combining the two positions that had been mandated by streamlining had been a mistake. Riutta had suspected as much when he had first arrived in Alameda, and saw how the district staff had been pared down to the point of near-paralysis, and 9/11 only served to confirm his feelings. Within two days, Riutta reorganized the district and the Area staff to fix the problem. The Area force was broken out and organized as an "N1-5" staff similar to a Navy command (with "O" and "M" combined in N-3), while the district staff was placed under the Reserve admiral in charge. Admiral O'Donnell took charge of California, while the N staff worked Area-wide problems and intelligence. The Crisis Action Center (CAC) was maintained briefly as a battle staff, and eventually subsumed as a battle staff by the Area "N" staff.

> The Incident Command System (ICS) worked well in the ports, where you had all the players, as in L.A./Long Beach. They had Coast Guard, INS, Customs, port police, National Guard, whoever happened to be there were all part of their ICS structure. For us, what worked better was a battle staff concept. We didn't need the ICS structure at the Area Command level. We weren't commanding an individual incident. We were managing a series of incidents, or potential incidents, all around a region.

As a commander, Riutta saw his role as not necessarily being able to walk into the room with all the answers, but rather with an attitude that conveyed the message that no problem was insoluble and let's figure it out. His responsibility then was to make sure that everybody "stayed in their lanes." The first time he walked into the

CAC it seemed like everybody in California, the district, the Area, and the MLC people, all wanted to be part of the action. Riutta reminded them that they would now split the work, because they were now at war and they all couldn't be there seven by twenty-four.

Riutta also sought to define the terms of reference. He looked for examples of leadership to Rear Admiral William T. Leahy, Commander of the Seventh Coast Guard District during Cuban and Haitian migration crises, and earlier Riutta's boss in Vietnam. Leahy's advice could be boiled down to, 'Take Action. Don't Wait!' He also reflected on the actions of James Loy as Area Commander during those same crises, in particular, Loy's ability to give his district commanders freedom of operation while Loy as Area Commander did battle with the bureaucracy to secure additional resources.

Riutta felt it important that his staff hear it from him that a war was on. For the foreseeable future, they had two missions: port security, and search and rescue. Everything else would be picked back up as time and resources permitted. The enemy has brought the battle to us, he told them, so we have to respond accordingly and fight this like a battle. Focus on the battle at hand, and we'll worry about all the other missions later. Secure the ports, keep commerce moving, and then we'll see where we go from there.

Unlike Richard Bennis in New York and Thad Allen in Portsmouth, Admiral Riutta was not faced with a massive response effort, which allowed him the time to plan for future incidents in the hope of blocking them before they happened. In this, Riutta took extra steps to increase the presence of the Coast Guard throughout the Area. Whether it was helicopters in the air over the ports, small boats in harbors, patrol boats along the coast, he considered it vital to keep Coast Guard assets in front of the public. He lost count of the number of times people would stop him and express their thanks at that presence.

For Riutta, such reassurance went to the heart of his mission as Area Commander—to keep commerce moving and the economy from imploding. "The first group of Sea Marshals we put on a cruise ship were cheered as they made their way ashore at Fisherman's

Wharf in San Francisco. When we had to start cutting back that program, the pilots would tell us how naked they felt without the Sea Marshals alongside them in the pilothouse." It all came together for the Admiral when he received a letter from a shopkeeper along the waterfront in San Francisco, who wrote to tell of the Coast Guard patrol boat that went by at different times and in different directions each day. These patrols offered such reassurance that the shop owner's customers were starting to come back.

> That is an impact on the economy, as well as an impact on the psyche. It not only makes us feel a little bit better about what we're doing, but it makes the public a lot more secure. That was a huge mission: restoring our faith in ourselves. When you bring a large cutter and anchor it in the middle of a bay, surrounded by small boats and with helicopters flying overhead, all of a sudden people say, 'okay, they have this under control. We can go about our business.'

For such reassurance to be offered the American people in the future, Riutta believed that the culture of the war-fighter within the Coast Guard needed attention. In the wake of the U.S.S. *Cole* incident, force protection and anti-terrorism had begun to reassert themselves as concerns for planners at both PACAREA and at Headquarters. But such pockets of study were scattered. Unlike the U.S. Army, with its staff and command schools, where mid-grade officers thought through and conducted war-gaming studies and exercises, the focus of Coast Guard post-graduate opportunities was on non-war-fighting specialties, business administration and civil engineering and the like. But after 9/11, Riutta believed, a gaping hole faced the Coast Guard officer corps—the lack of a command and staff college, a war-fighting school, that would serve through historical studies to prevent the kind of unilateral disarmament of the stations and the dissolution of port security units that had taken place after the fall of the Berlin Wall.

> It was too expensive, nobody's going to attack us, peace in our time, life is good. Every generation goes through that mistake. Now that we've had our rude awakening and we're not going to make that mistake for another twenty years. If nothing happens between now and then, we'll go

through this whole situation again. Hopefully there will be some warfighters left to say, 'Hold it guys, you've got to go back.'

Unfortunately, in the new world disorder, a country that went to sleep for even one day risked millions of its citizens to a weapon of mass destruction. For Riutta, the key to all planning was intelligence. Without the ability to defend everything, intelligence—hard intelligence that was collected, fused, and disseminated to operational commanders who could match it with local intelligence to create a joint or multi-agency threat assessment—was the new element that would leverage all other Coast Guard operations. In a Service that historically spread itself across a wide mission spectrum, it remained to be seen what level of effort and dollars the senior leadership devoted to this singular task, without which all other economic security missions were on course for failure.

Chapter Five

9/11 and the Office of the Commandant: James Loy as Strategic Field Commander

If you walk down the hall outside the offices of the Coast Guard's Atlantic Area Commander in Portsmouth, Virginia, you pass a series of paintings of all of the three-star admirals who have held that position. There is Paul Welling, smiling, hands on hips, on the bridge of a cutter, conveying optimism and energy, and Howard Thorsen, in front of a faded American flag conveying both patriotism and sense of history. Then you come to a portrait that stops you dead in your tracks. It shows a three-star admiral in dress whites, on the bridge of a ship—impossible to tell what type—with a glass in his hand and storm clouds gathering over his broad shoulders. The look in his eyes conveys both authority and wariness, and were it not for the many-tiered rack of modern naval ribbons, you would be hard pressed to tell whether this officer, James M. Loy, had commanded a naval vessel from the year 1995, or the year 1795.

James Loy was a field commander risen to the rank of Admiral. For two years he oversaw the Coast Guard's participation in Operation *Uphold Democracy*, a U.S.-led multinational military intervention to restore the democratically-elected government of Haiti, combined with the mass Cuban and Haitian migrations of late 1994 and early 1995. All this while shifting Atlantic Area operations from its traditional home on Governor's Island in New York Harbor, to the new residence of LANTAREA staff in Portsmouth. By implication, the move signaled a new and closer operational relationship with the U.S. Navy, not only in overseas expeditionary maneuvers, but in Loy's developing doctrine of the Coast Guard as "a unique instrument of national security" and his growing sense that in the post-Cold War world, the United States faced new and unpredictable "asymmetrical" threats from around the globe.

The storm clouds in the background of the painting were prophetic. Loy would advance from his command of the Atlantic Area to become Chief of Staff at Headquarters in Washington, D.C., at a critical juncture in the Service's history. He spent two years breaking down and rebuilding the planning, programming, and budgeting system for the Service, which had become encrusted with laborious and ultimately futile staff work that held little hope of surviving the annual budget process. Loy's ideas—for example to get the direct input of operational field commanders into the budget process—were then taken by Vice Admiral Timothy W. Josiah, who succeeded Loy as Chief of Staff and served in that position for four years, and transformed into a new and efficient process for developing the famously precarious Coast Guard operating budget.

After two years as Chief of Staff, Loy received his fourth star, and became the 21st Commandant of the U.S. Coast Guard, at a time when each fiscal year seemed to bring new threats to the very survival of the Service. The Coast Guard had responsibility for the security of American ports. But since the end of the Cold War—and the end of the threat of a general war in Europe—such security concerns had been allowed to fade into the background. The Coast Guard Reserve, upon whom the burden of port security rested, had been allowed to shrink through Congressional budgetary neglect by more than one third, from 12,500 sailors in 1988 to little more than 7,500 by the time Loy took command. An across-the-board twelve per cent "streamlining" process for active duty forces, initiated as an aggressive compliance to government-wide reductions in the early 1990s, had gutted four hundred million dollars from the annual operating expenses budget, and led to the loss of 4,000 active duty personnel, nearly ten per cent of the entire active duty roster.

Such cuts fell hard throughout the Service. The Coast Guard's large work platforms, its sea-going cutters, patrol aircraft, and rescue helicopters, operated for a set number of budgeted hours per year. When they were at sea, structural failures illustrated the recycled nature of the Service's aging fleet, none so dramatic as when the small boat davits on the cutter *Storis*, a veteran of the Second World War, gave way in the middle of the Bering Sea, dumping nine sailors into icy Arctic waters. All were recovered, numb from hypothermia.

Following hard on the heels of streamlining, Loy arrived at the Commandant's Office in mid-1998 with barely a year to ready the Service's computers for Y2K. The budget for acquisitions, construction, and improvements was cut nearly in half from fiscal year 2000 to fiscal year 2001. An additional across the board cut in the Coast Guard's budget in the fiscal year prior to 9/11 had further demoralized an already diminished cadre, at the very time when all thoughtful personnel within the Service knew that their operational challenges were increasing each day. Loy could be forgiven for leaving his office at night and pausing to commiserate with the portrait of another titanic Coast Guard Commandant, Russell Waesche, the man who led the Service through the Great Depression and the Second World War. As with Waesche, Loy is credited in many quarters with saving the Service from disaster in the midst of a global crisis.

Loy had several fixed stars by which to navigate this course. Alexander Hamilton's notion of a federal marine agency that would act as sentinels of the sea gave the Coast Guard a clear national strategic mission that could be traced back to the Founding Fathers. The first American lighthouse, built on Little Brewster Island in Boston Harbor in 1716, imparted to the Service a marine safety mission older than the country itself. The advance of maritime technologies and the global movement of economically displaced peoples throughout the 19th and 20th centuries both served to pile further missions upon the Service. By the time Loy took office, the U.S. Coast Guard gathered the authority for its many missions from over 3,500 separate laws, rules, regulations, treaties, and executive orders.

Loy was assisted mightily in his crusade to re-ballast the Coast Guard by a 1999 Office of Management and Budget (OMB) report that supported in every particular his contentions about the intensifying value of the Coast Guard's roles and missions in a changing world. The report came at a time when the Coast Guard was attempting to rally congressional support for its planned "Deepwater" acquisition project, a multi-year, multi-billion dollar effort led at the time by Rear Admiral Roy Casto that envisioned not merely replacing cutters like the *Storis*, but rethinking the whole fusion of intelligence, computer systems, operational platforms, and people.

Looking twenty years into the future, the OMB report concluded that if the U.S. did not have a Coast Guard it would have to invent one. Each mission of the Coast Guard was studied with the result that each was predicted to increase in importance to the nation. The value of the Coast Guard to the nation was seen in its realization of each of these necessary missions under the umbrella of a single organization, with a single overhead cost to the federal treasury.

The OMB study recommended that the Coast Guard employ state of the art, off the shelf technologies. This strategy was applied to both the modernization of the National Distress and Response System and the Deepwater Project, both projects forming the core of Loy's attempt to rebuild the Coast Guard. OMB supported both the rationale for Deepwater as well as the Coast Guard's plan to make it happen.

The sixth conclusion of the OMB report focused on the flexibility of the Coast Guard in quickly responding to national emergencies. It was this singular quality that OMB implored the Coast Guard to value almost more than any other. Nowhere was this notion put to greater test, for both the Coast Guard and its Commandant, than on the morning of 9/11.

On the morning of 9/11, the Admiral was in his office at Coast Guard Headquarters in Washington, D.C. He had just concluded a meeting with Admiral Stillman and several other people associated with oversight of the Deepwater Project, when a staffer appeared and said he might want to turn on his television. Seeing the images coming from Manhattan, Loy's mind went immediately to terrorism.

The impact of the second aircraft confirmed his fears. He rallied his top staff deputies: the Vice Commandant, Vice Admiral Thomas H. Collins, and the Chief of Staff, Vice Admiral Timothy Josiah, as well as the heads of the two main branches of the Coast Guard operational tree: Rear Admiral Terry M. Cross, Assistant Commandant for Operations, and Mr. Jeff High, the civilian Director of Waterways Management (G-MW) (the deputy of Rear Admiral Paul J. Pluta, Assistant Commandant for Marine Safety and Environmental Protection, who was away at a conference in China) to join him in thinking through the Service's response. Captain Anthony Regalbuto, Chief, Office of Policy and Planning (G-MWP) within the Waterways Management Directorate (all "M" port security responsibilities were within his office) also attended many of the meetings and led the Command Center Incident Management Team activation for the first 24 hours after the attacks.

As he gathered his staff, Loy made two quick phone calls. Within seconds of the second plane's impact, Loy knew he needed his Reserve. But first he needed to find Norman Y. Mineta, the Secretary of Transportation and Loy's civilian boss, who held the power under Title 14 of the U.S. Code to mobilize the Coast Guard Reserve in the face of a domestic emergency. Mineta, at that moment, was being pulled from his office and taken to secure spaces within the White House, where in the absence of the President, the Vice President had gathered the top administration staff. Mineta's cabinet post oversaw several government agencies besides the Coast Guard, and one of them was the Federal Aviation Administration. It was to deal with the aviation crisis that Mineta was called to Pennsylvania Avenue. Loy got through telephonically to the White House, and within minutes of the start of the crisis received Mineta's blessing for a call-up of 5,000 Coast Guard Reserves. A few moments later, an equally dramatic phone call came from Chief of Naval Operations Admiral Vern Clarke, asking what the U.S. Navy could do to assist the U.S. Coast Guard.

For Vice Admiral Josiah, the days and weeks ahead would reveal the thinness of both the Coast Guard's bench and its budget. The fiscal year 2001 budget had left the Service ninety million dollars short of paying its bills and its people. Normal operations had increased to such an extent that, for example, 110' patrol boats were being run 2,500 hours a year rather than their planned 1,800. In the years prior to 9/11, the budget for fuel alone was annually running tens of millions of dollars short of program needs. An armed Service, the Coast Guard was wedged into a federal department where it competed for operational dollars with Amtrak, highways, and the federal air traffic control system. Many thought the Service would be better off in the Department of Defense, or Justice. Josiah himself had commissioned a study that concluded that the Coast Guard would be better off as an independent agency.

With Reserve strength on 9/11 down more than a third over what it had been for *Desert Sheild/Desert Storm* a decade earlier, and most of the remaining Reservists operating to augment the everyday active duty force, the Service had little surge capability to bring to the fight. "We didn't have the depth to do something that was a monumental tasking," recalled Josiah, "that wasn't going to go away, that put all of our people under twenty-four hour stress, and for the first time in a long time led the nation to recognize that the coast was 95,000 miles long, with 361 ports containing thousands of waterfront facilities. All of a sudden the country wanted all of that protected from the waterside."

For Loy, the attacks unfolded almost in slow motion, the result of his years of warning that such an attack, an "asymmetric" attack using unconventional weapons, was not only possible but likely. Now the disaster that he feared was upon him, and the years of intellectual study he had invested in unconventional terrorist scenarios, and in both building and promoting the U.S. Coast Guard as a unique instrument of national security, allowed him to do the bureaucratic equivalent of some open field running.

As Loy was quick to point out to his counterparts in the U.S. Navy, most navies of the world far more closely resembled the U.S. Coast Guard than the U.S. Navy. Like most foreign navies, the U.S. Coast Guard existed in a multi-mission universe, a universe of continuous operations both civilian and military in nature. That experience in part led Loy and his staff to immediately ask themselves: 'if they are hitting the World Trade Center today, where might they strike tonight, much less tomorrow.' As it turned out, the terrorists struck their next target little more than an hour after Loy mobilized the Reserve.

Loy is very likely the last Coast Guard Commandant who will have started his career by seeing combat on patrol in Vietnam, in Loy's case as commander of the 81' patrol boat *Point Lomas*, and he carried close within himself the need to impart his direct personal combat experience to those around him. It was the reason he sprinkled so many of his speeches with history and historical anecdotes. He knew that the generation of Coast Guard leaders coming up behind him needed that perspective, and he was determined that they would get a large measure of it from the boss.

> Combat experience is an amazingly maturing process for a young person, certainly for a young officer. I went over there responsible for those twelve guys who became blood brothers to me. You find out very quickly your personal strengths and your personal weaknesses.

But he was not the type to strike the pose of grizzled war veteran to make his point. For twenty years he had held his Vietnam experience at arm's length, in large measure because of the general rejection of Vietnam veterans when they first came back and, as Loy saw it, "the discoloration of patriotism and harmony that had always been part of our national experience in that regard in the past." During his post-Vietnam stay at Wesleyan University, while preparing for a teaching position at the Coast Guard Academy, Loy and several other veterans of the war decided to wear their uniforms on Armed Forces Day. The subsequent ostracizing by those who had previously been friends and colleagues remains vivid in his mind to this day. Like so many of his comrades, Loy for years had endured the doubled stress of dealing with what he had seen overseas combined with his attempts to fit back into a society that had rejected his service.

On the morning of 9/11, Loy's press aide, Captain Mike Lapinski, a former buoy tender and ice breaking tug commander, was in his office doing some staff work when the attacks commenced. He made his way down to the boss' office, following his perhaps natural instinct as a Coast Guard officer that in almost every instance an initial catastrophe would get worse very quickly. He had just entered Admiral Loy's office at 0943 when a boom rattled the windows at Coast Guard Headquarters. Something had been hit locally. From his time in Vietnam, Loy recognized an explosion when he heard one. Loy looked up and said, "My God, that's too close." Lapinski saw the Admiral suddenly transformed into his once-familiar role of combat commander. The "twelve guys on an 81-footer" now became 40,000 active duty sailors, 7,500 Reserves, 33,000 Auxiliarists, and several thousand civilians.

Master Chief Petty Officer of the Coast Guard Vince Patton was no stranger to either Coast Guard history or to the military valor embedded within it. For twelve years he had worn a silver bracelet inscribed with the name of a Coast Guard aviator: Viet Nam MIA, Lieutenant Jack C. Rittichier, who was piloting an Air Force "Jolly Green" helicopter, attempting to rescue a downed pilot, when the helo came under enemy fire, crashed, and exploded. Patton wore the bracelet as a symbol both of the Coast Guard as a military Service as a reminder for others in the Service not to forget the Coast Guard sacrifice there.

One of five most senior enlisted personnel in the U.S. Armed Forces, the Master Chief counted it as part of his portfolio to keep an eye on the old man. As the scope of the attacks became clear, and he felt his office shake from the explosion at the Pentagon, Patton went the short distance from his office to where the Commandant and his senior staff were receiving reports from around the country

> Admiral Loy was the coolest, calmest, most calculating individual that morning. And when he realized that nobody had their stuff together, he just started rattling off what had to be done, how it had to be done, let's do this, let's do that. He just put it all together. Everyone was running around trying to get him factual information, and when he realized that that wasn't coming, he grasped the situation so quickly, and began to formulate the Coast Guard plan off the top of his head. From that point for-

ward, my attention was all on him. I can't tell you who else was in that room, but I can tell you that everybody else sure knew that he was there. I actually left that room with tears in my eyes, thinking that, whatever else comes out of this situation, the Coast Guard is in good hands.

As Patton's comments imply, Loy may have been surprised at the proximity of the attack but the attack itself was hardly a shock. In a remarkably prescient article for the *Homeland Defense Journal* co-authored before the 9/11 attacks with Captain Robert G. Ross, Loy had argued that rather than provide a new era of peace and security, the end of the Cold War had perversely done just the opposite:

> In this era of globalization, the world reach of America's economy and culture are creating powerful resentments in some sectors. Even without regional conflicts providing motivation, it is highly likely that Usama bin Laden or a similarly reactionary guardian of traditional ways would have arisen in reaction to the dominance of modern America's economic power and culture, some aspects of which are admittedly negative. Those with a dislike for the effects of globalization, as well as those who merely feel threatened or left out by an economy and technology they do not understand, have strong motives for lashing out at the most highly visible source of their discontent: the United States.

Given the Coast Guard's multiple responsibilities for the economic security of the United States, Loy's linkage of the globalization of American culture with reactionary fundamentalist terrorism was not an idle one. In his college days at Wesleyan he had read Charles Beard's *An Economic Interpretation of the Constitution of the United States*, and its thesis that American power emanated as much from superpower economics as the Bill of Rights, left a lasting impression. Loy understood better than most that, hand in hand with mass murder, a reactionary terrorist like bin Laden sought general economic disruption and destruction. Loy had to head him off at the port, by balancing security with access. Anyone, given enough operational platforms, could lock down the ports of the United States. But this played directly into the hands of the enemy. Loy's challenge was to secure the ports while at the same time maintaining the free flow of goods essential to a global economy.

The entire Coast Guard was now at war, and with a depleted Service Jim Loy was responsible for coordinating the suddenly-threatened security of over 350 American ports, 12,383 miles of coastline, 88,633 miles of shoreline, and 26,000 miles of strategic and economically essential navigable channel waterways. Loy had to do all this while fully engaged in his two primary concerns before 9/11. The first was his ongoing effort to ratchet down the increasing tempo of the Service's other day-to-day operations. These included search and rescue, fisheries and other economic security patrols, counter-drug and migrant interdiction operations, polar ice-breaking and support of national science objectives, and the environmental protection of the American waterfront and maritime frontier. The second was his engagement with the future, with his vision of what the Coast Guard would look like in the year 2020, and development of the people, resources, and technologies required to begin what will become a total transformation of the Service.

As the morning advanced, Loy and the staff at Coast Guard Headquarters absorbed a series of reports—all but one of which proved false—about continuing general attacks on Washington, D.C. A plane had landed on the Mall, a car bomb had exploded outside the State Department, a fourth airliner had crashed, this time in Pennsylvania, apparently while on course for a target in the district. What was clear was that the Coast Guard's Atlantic Area was under attack. Loy got both of his Area Commanders on the phone and pressed Vice Admiral Ernest R. Riutta in Alameda, California, and Vice Admiral Thad Allen in Portsmouth, to have faith in their local field commanders.

> The first thing I said to the Area Commanders was to understand that we don't know what might be next, but your challenge is to have faith in those local Captains of the Port and the local district commanders, and as necessary the next level down, the group commanders. And the first step in that process is to focus, in the classic defense nomenclature, on the 4 shop rather than the 3 or 5 shops, focus on the logistics apparatus which will be vital to supporting those Captains of the Port. As long as you have communications with your district commanders, find out what

they need and get it to them as quickly as possible. We'll notionalize the big picture in due time, but as long as the crisis is in front of us let's try to define what might be next and defend against it.

The ports and waterways had to be defended, and it would take a Coast Guard-wide effort to do so. The long-standing separation of the Marine Safety and the Operational sides of the Coast Guard—the cultural differences that differentiated the regulators from the boat and aircraft drivers—had in recent years begun to draw closer together, both at the local level and as a matter of Service-wide policy. 9/11 would prove to be the event that cemented this shift once and for all. Operational platforms like medium endurance cutters suddenly appeared within major U.S. ports, giving the Captains of those ports a secure command and control platform, as well as offering reassurance to a shaken public.

The spike in the operational hours on those platforms, to say nothing of the cost, became an immediate concern. The Coast Guard spent about 67 cents of every dollar on its people, leaving less than a third of its budget for operations, maintenance, and logistics. The increased tempo of operations required more gas and spare parts, and there was little left in the budget for either.

The spike led to immediate concerns on other fronts as well. For example, on the West Coast, Vice Admiral Riutta called the Pacific fleet toward home waters, to secure the four fundamental U.S. Pacific ports of Puget Sound, San Francisco, Los Angeles/Long Beach, and San Diego. The U.S. Navy had gone from its lowest force protection condition "Alpha" on September 10th, to condition "Charlie," and in some ports, to the highest level "Delta." As the nation's fifth armed Service, the Coast Guard aligned its force protection with that of the Navy. It quickly became apparent, however, that the Coast Guard lacked the deepwater assets, the numbers of cutters and aircraft, to both protect the ports and carry out all its other responsibilities.

Loy insisted that the Service match its entire inventory, small boats, Marine Safety Offices, cutters and aircraft, all of it, against the Navy's current force protection doctrine, with the result that the

Coast Guard generated, in effect, a set of force protection conditions of its own. This new maritime security, or MARSEC, doctrine involved three distinct conditions. The first, MARSEC ONE, aligned the Coast Guard with the Navy's force protection "Bravo" level, and allows, in effect, the major cutter fleet to go back to work offshore. Should intelligence reports indicate a need to kick security up to MARSEC TWO, then the Service would initiate steps to borrow assets from the deepwater fleet for homeland maritime security operations. The third and highest level, MARSEC THREE, would indicate an imminent attack.

Once this new maritime security doctrine was in place, Loy was able to project the array of forces at the Service's command at any given threat level. With that force lay-down in effect, Loy then projected a three-year budget build, commencing in Fiscal Year 2003, to give the Coast Guard the assets it required to retain all its mission capabilities while at MARSEC ONE, and then having the flexibility to adjust assets into levels TWO and THREE as necessary.

As for the ports themselves, Admiral Loy understood that American maritime ports were qualitatively different from American airports, highways, or railways. Road and rail systems, and especially the air traffic system, were only enabled through the investment of massive federal subsidies in the late 19th century and throughout the 20th century. Ports, on the other hand, came down to us from the 1600s and 1700s by and large as privately financed economic infrastructures. Any new federal role in securing actual port structures and facilities would be a radical departure from the historic American experience on the waterfront. But such a regulatory and program shift would be nothing new to the Coast Guard.

With the formation of the modern Coast Guard in 1914, the first Commandant, Ellsworth P. Bertholf, had insisted that the Service retain its multi-mission nature. The character of those missions would change over time, and with them budgets and personnel and technologies. New missions would often be added by Congress and the Executive Branch without the accompanying personnel or budgets to perform them. The modest Coast Guard of the 1930s grew exponentially to the 250,000 Coast Guard personnel who fought in the Second World War. The wreck of the *Exxon Valdez* in Alaska led to the Oil Pollution Act of 1990 (OPA-90), and to more than one hundred new regulatory projects in the ensuing decade, and a shift of the Coast Guard towards the prevention of and response to major oil spills. A new administration in 1992 had asked all government departments to offer up ten per cent of their operations to streamlining. In a disastrous downshift, the Coast Guard offered up more than twelve per cent of its workforce even as the new programs and regulations that spun out from the *Exxon Valdez'* oil slick spread across the Service.

As Chief of Staff Vice Admiral Josiah saw it, the Service was built around response. Whenever anyone called 9-1-1 in the maritime domain, the Coast Guard was always there. "We do some patrolling, but now [after 9/11] the need was to be out there seven days a week, twenty-four hours a day, and be planted near some of the nation's critical port and waterways facilities. It was clearly more than we could do."

What Admiral Loy confronted on 9/11 was the biggest port security operation since the Second World War, and he met this challenge with a Service that had dwindled to less than one-seventh the size of the force that Admiral Waesche had at his command in 1945. Loy knew he would never command those kinds of resources, either in terms of personnel or cutters, so he tapped into one of the core strengths of the Coast Guard, its ability to prioritize missions and attack those with the highest threats to national security and safety at sea. Combined with outreach to local port authorities and harbor safety committees, who would have to see to the security regimes of their own cargo terminals, to cruise lines responsible for the security of the thousands of passengers and tons of baggage they embarked, Loy would have the "all hands evolution" he required to strengthen the security of American ports and waterways.

The Coast Guard's immediate role would be to secure the ports, increase the amount of time the Coast Guard had to examine inbound commercial vessels from twenty-four to ninety-six hours, and then conduct an intensive round of Port Vulnerability Assessments (PVA), which would identify the key areas of each port

and which player, private enterprise, or local, state, or federal entities, would have responsibility and authority for each of those areas.

> If 9/11 showed anything about us as an organization, it was that the crisis of the moment could be reached by our Service doing what it does best. I literally was able to pick up the telephone and tell the field commanders: 'Take a left and go to port security.' And all those people who were doing all those other things, we went from having 2% of our budgeted capability going to port security on 9/10, to almost 58% in a matter of weeks, an astonishing ability as an organization to shift gears and go where the nation needs us.

Getting this word out to the public, to say nothing of the fleet and the units that supported it, proved difficult at first. The Department of Transportation had sent out directives early on in the crisis, ordering the Coast Guard essentially to say nothing about its operations. Rear Admiral Kevin J. Eldridge, Assistant Commandant for Government and Public Affairs, and his savvy Chief of Coast Guard Public Affairs, Captain Jeffrey Karonis recognized immediately that such a *de facto* gag order was self-defeating. In any crisis, the Coast Guard needed to drive its own coverage, and not allow events to drive the coverage about the Service. "We have thousands of people out on the water, on ships and boats, and they needed to know what was going on," remembered Karonis. "This wasn't something you could just run and hide from without disclosing anything. Our second concern was reassuring the public. If you can't talk, you can't reassure the public." Karonis also asked for permission to send a cameraman to record the actions of the senior leadership in a crisis, something that had never been done before in the Service's history. Admiral Loy agreed immediately. Admiral Eldridge later approved a project to document Coast Guard operations on 9/11, the first such effort at official Service history since 1946.

Loy recognized as well the necessity of communicating the role of the Coast Guard both publicly and privately. Soon after 9/11, Admiral Loy went to New York to thank his people for their service, and his time at Ground Zero and with the Coast Guard personnel in and around New York Harbor became a spiritual experience. He saw the same looks of combat shock and fatigue he remembered from Vietnam, with an important difference. "Vietnam was long periods of almost boredom interspersed with moments of sheer panic. 9/11, and for weeks thereafter, was a constant, unrelenting requirement for service, at all levels of our Service, humanitarian, operational, military."

The trip was difficult personally as well. Master Chief Patton had served with Loy when the Coast Guard presence on Governor's Island came to an end. He remembered that as the Coast Guard flag was lowered during closing ceremonies, that it was the first time he'd seen the boss get emotional. But any anger Loy may have felt after 9/11 he found himself channeling into the management of its consequences, and moving as quickly as possible to prevention of future attacks.

For Loy, 9/11 showed that the Coast Guard was still the nation's premiere response agency. No other federal agency existed that could shift its focus as instantly, and bring people and platforms to bear on a disaster of national scope. If there was fault to be found, it was Loy's view that during the decade of the 90s the country had lost the edge in intelligence—human, signal, electronic—that had always existed during the Cold War, capabilities that might have acted to trigger the early warnings of attack that had been the indispensable element in the prevention of a nuclear war.

For the Coast Guard, the new world required that it develop what came to be called Maritime Domain Awareness (MDA), a sweeping and continuous view over the maritime horizon to detect potential threats long before they became attacks on the maritime frontier or, worse, in a strategic port itself. For Loy, nineteen guys without guns had been able to carry out their infamy on 9/11 in large measure because the awareness piece of national security had lapsed. Exactly what immediate role the Coast Guard would play in this regard was very much up in the air. It was a confusion of roles that the Chief of Staff, for one, very quickly found disquieting.

Vice Admiral Josiah remembered looking out his window the day after 9/11 to see an 87' Marine Protector Class patrol boat cruising along the confluence of the Anacostia and Potomac rivers. "At first I thought, 'That's pretty comforting.' Then I started to ask 'What is it doing? What is its' tasking? Is it empowered to use force?'" Josiah went so far as to hitch a ride out to the patrol boat, to thank the crew for their work, and also to gain a more direct sense of how Coast Guard vessels were being tasked, deployed, and operated to defend national assets within line of sight of the Potomac. What he found was a less than reassuring lack of plans, cooperation, and communication between various waterfront security regimes, marine police, local police departments, and the Coast Guard patrol boat, which was a long way from its normal area of responsibility (AOR) and had never patrolled the river before.

To begin to sort through these increasing complex burdens on the Service, Admiral Loy went back to his native Pennsylvania soon after the President had appointed Governor Tom Ridge of that state to become the White House's Director of Homeland Security, to meet with him before he reported to Washington. He knew that a barrage of jobs that would inevitably be dumped in Ridge's basket, and suggested that the Coast Guard could take out the maritime security component, work the problem, and bring it back to Ridge fully developed. The reorganization of federal border security apparatus had long been an interest for study groups, and the Coast Guard seemed to figure at the forefront of all such studies, as if it could be pulled, like an electric plug, and seamlessly reinserted into a different federal power socket on another side of the bureaucracy.

It had happened in 1967, when President Lyndon B. Johnson extracted the Coast Guard from the Treasury Department and placed it into a new Department of Transportation, in effect to provide organization ballast to a cabinet post that otherwise would have very little. As late as April of 2001, the Hart-Rudman Commission had recommended the consolidation of the Coast Guard with FEMA, the Border Patrol, Immigration and Naturalization Service, and Customs, and several others, into a kind of super border patrol service. The Coast Guard now entered anew into every discussion of federal reorganization in the face of international terrorism and the defense of the homeland. But many of the discussions, which picked and chose Coast Guard missions for the new agency like dishes at a buffet, missed the mark.

For Loy, the important lessons for the American people to take from the Coast Guard's performance on 9/11, is that the shift into the port security mission could just as easily have been a shift into a Cuban migration crisis, an increase in counter-drug law enforcement, a major fisheries challenge in the Bering Sea or on the Grand Banks, or an oil spill. It was the organic whole, supported by a unified training regime under a single overhead that allowed the Coast Guard to survive cuts in its budget, losses of personnel, and deteriorating equipment. Had Loy possessed on 9/11 the 4,000 active duty bodies that had been lost to streamlining, he believes he could have done much more, with less reliance on calling up the reserves, than he was able to do.

The efficient performance of the Reserve and Auxiliary after 9/11 tended to hide just how exposed the ports were and how thin the active duty Service had been spread. But the consequences of that exposure were to reverse a decade of reductions in the Coast Guard's work force, as Loy sought and won back in the post-9/11 budget cycles more than half the 4,000 active duty personnel lost to streamlining, along with a simultaneous commitment to rebuild the Coast Guard Reserve back to its pre-streamlining force of 12,500.

When combined with the subsequent approval of a contractor to oversee the development of the Deepwater Project along with the budget commitments to make it happen, the 9/11 performance of Jim Loy was certain to fix him as one of the greatest commandants in the history of the Service. The storm clouds embedded in his official portrait in Portsmouth had intensified into the greatest storm the Service had weathered since the Second World War. Spy glass in hand, Loy had laid down the trackline for the Service's way out of the storm and into the future.

Almost as Loy was walking out the door, in May, 2002, the White House announced that the Coast Guard would indeed move, from Transportation to a new Department of Homeland Security. Just how and when this move would be effected, or how the Service would build an entirely new port security doctrine, or how the 3,500 rules, laws and regulations that governed the Coast Guard would be unplugged from Transportation and transferred to a new cabinet secretary, would have to be left to Loy's successors.

Top: Boatswain's Mate Third Class Carlos Perez (center) piloted the first Coast Guard boat to respond to smoke coming from the World Trade Center. (Photo by Petty Officer Michael Hvozda)

Right: Rear Admiral Jeffrey Hathaway. Hathaway was in charge of the U.S. Navy Command Center at the Pentagon when it was destroyed on 9/11. (Photo courtesy of U.S. Navy)

Above: An aerial view of the still smoldering World Trade Center Complex September 26, 15 days after terrorists attacked the twin towers with hijacked jetliners. (Photo by Chief Petty Officer Brandon Brewer)

Below: Seen from U.S. Coast Guard Headquarters, clouds of smoke billow from the Pentagon after the attack that destroyed the U.S. Navy Command Center. (Photo by Telfair H. Brown)

Above: Admiral James M. Loy gathers many of his top deputies on September 12, 2001. From left: Admiral Loy, Vice Admiral Timothy W. Josiah, Captain Anthony Regalbuto, Mr. Jeff High, Captain David Pekoskie, Captain John E. Crowley, Captain Robert Papp, and Rear Admiral Kevin J. Eldridge. (Photo by Garcia Graves)

Below: A week after the attacks, Admiral Loy and his senior officers work through the latest situation reports in the headquarters Command Center. Vice Admiral Josiah standing at left and Rear Admiral Eldridge standing at right. Seated from left are Admiral Loy, Vice Admiral Thomas H. Collins, Rear Admiral Terry M. Cross, Rear Admiral Ken Venuto and, back to the camera, Rear Admiral Paul Pluta and Mr. Jeff High. (Photo by Garcia Graves)

Left: Vice Admiral Thad W. Allen, Commander of the Coast Guard's Atlantic Area on 9/11, diverted every major cutter underway that morning to a port. To ensure command and control, "cutters were the immediate solution." (Photo by Petty Officer Patrick Montgomery)

Below: Vice Admiral Ernest Ray Riutta (right), Commander of the Coast Guard's Pacific Area on 9/11, immediately deployed his Reserve port security units to Long Beach and Seattle. Without the Reserve and Auxiliary, "we would have been dead in the water on September 12th." (Photo by Warrant Officer Lance Jones)

Above: Chief Warrant Officers Leo Deon (left) and Leonard Rich from the Coast Guard's Atlantic Strike Team discuss safety strategy at Ground Zero. (Photo by Petty Officer Tom Sperduto)

Right: The Master Chief Petty Officer of the Coast Guard, Vince Patton, reads some of the messages that were applied to the first responder fire truck near the World Trade Center disaster site. Patton was visiting with Coast Guard personnel assisting in the recovery efforts. (Photo by Petty Officer Mark Mackowiak)

Above: After seeing the attacks on television, Captain Richard P. Yatto, commander of Air Station Cape Cod, immediately recalled his helicopters in a dramatic bid to get them to New York to attempt a rooftop rescue before the twin towers collapsed. (Photo by Chief Petty Officer P.J. Capelotti)

Left: Emotions evident, Rear Admiral Richard Bennis, Commanding Officer of Activities New York, greets Admiral Loy upon the latter's arrival at Newark International Airport, September 23, 2001. (Photo by Petty Officer Tom Sperduto)

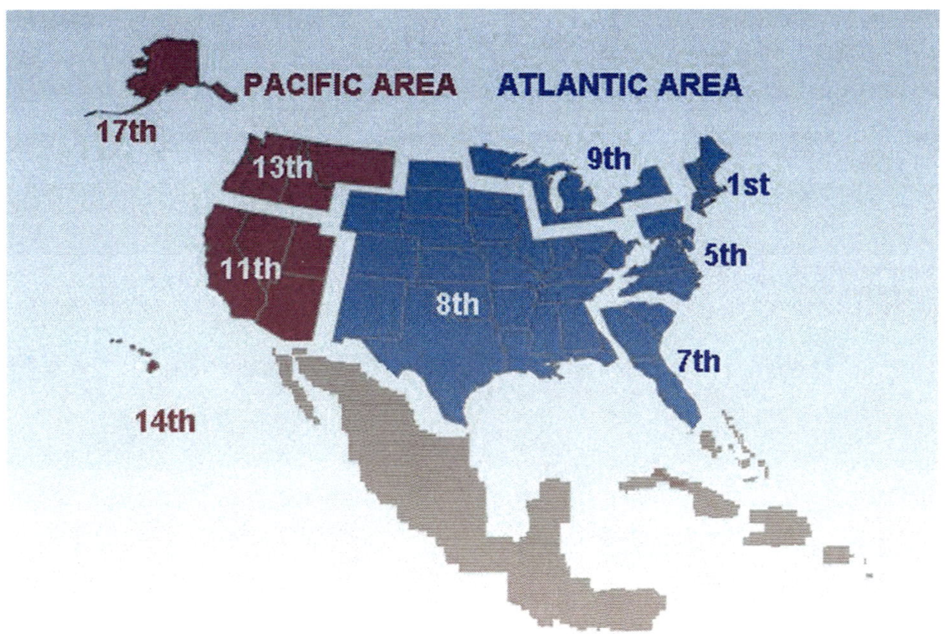

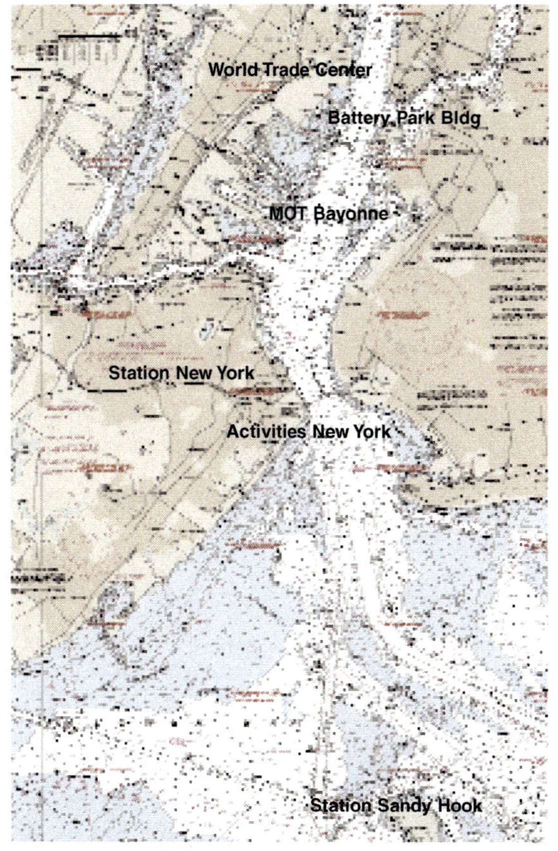

Above: On 9/11, the Coast Guard district's were under the command of the following senior officers: 1st: Rear Admiral George N. Naccara; 5th: Vice Admiral Thad W. Allen; 7th: Rear Admiral James S. Carmichael; 8th: Rear Admiral Roy J. Casto; 9th: Rear Admiral James D. Hull; 11th: Vice Admiral Ernest "Ray" Riutta; 13th: Rear Admiral Errol M. Brown; 14th: Rear Admiral Ralph D Utley; 17th: Rear Admiral Thomas J. Barrett.

Left: Coast Guard facilities around New York Harbor, September, 2001.

Above: A Coast Guard transportable small boat from Reserve Port Security Unit 305 and a Coast Guard helicopter patrol Guantanamo Bay, Cuba, during detainee operations in the spring of 2002. (Photo by Petty Officer Krystyna Hannum)

Left: The portrait of Admiral James Loy painted during his tenure as Atlantic Area Commander. Loy's intellect, vision, and resolution led the Coast Guard through 9/11 and into its lead role to secure the American maritime frontier. On the morning of 9/11, said Vince Patton, "I can't tell you who else was in that room, but I can tell you that everybody else sure knew that he was there."

Chapter Six

9/11 and the Coast Guard at the Pentagon: Jeffrey Hathaway and the Navy Command Center

Unknown to most people even within the Service, a rotation had existed for some time wherein a Coast Guard flag officer served as part of the U.S. Navy operations staff at the Pentagon. The job was not a liaison position but rather filled a Navy flag billet, offering a Coast Guard flag officer experience in Navy operations while raising the visibility of the Coast Guard among the senior Navy staff. Historically, the position involved exposure to daily U.S. Navy operations, including oversight of the Navy's Pentagon Command Center, which was the Chief of Naval Operations link to the Fleet. Amongst other responsibilities, the Coast Guard admiral kept track of Fleet movements for the Navy.

The admiral selected for this assignment needed an extensive operational background, so it always went to either a sea-going officer or an aviator, one newly promoted to the one-star rank. In the years prior to 9/11, the position had been filled by Rear Admiral James Carmichael, a cutter captain who on 9/11 was Commander of the Seventh Coast Guard District in Miami, and most recently by Rear Admiral Ralph Utley, an aviator who had left the Navy position in the Pentagon to assume command of the Fourteenth District in Honolulu in June of 2001. Utley was replaced in the Pentagon billet by Rear Admiral Jeffrey Hathaway.

When Hathaway arrived at the Pentagon in the summer of 2001 to relieve Admiral Utley, the N3/N5 staff of the Navy, operations and plans and policies, was contemplating a reorganization that would place a flag officer uniquely in charge of N34, the Navy's director of anti-terrorism and force protection (AT/FP). As part of this reorganization, Admiral Hathaway volunteered to take over N34, as well as elements of N32, the Navy's support to interagency anti-drug efforts. The Command Center, it had been decided, would be split off from Hathaway's responsibilities. To complicate this organizational shift, the Command Center had also just finished a physical move, from old spaces within the Pentagon to the first deck of the newly-renovated wedge of the building. Also physically located in the Command Center were Hathaway's interagency counter-drug (N32) people.

Hathaway had been in charge of N32 and N34 policies for about two months, while still retaining oversight of the Navy Command Center, which had not yet been split off from his responsibilities, when he arrived at the Command Center inside the Pentagon at his usual time of 0545 on the morning of 9/11. For all intents and purposes it was a normal morning. Hathaway gathered status reports from around the globe, in preparation for the 0700 brief to Vice Admiral Keating, the three-star in charge of N3/N5, who in turn would brief the Chief of Naval Operations later in the morning. Hathaway then put folks in the Command Center on task, prior to leaving the Pentagon at 0800 for Coast Guard Headquarters. There he was scheduled to head up a school selection board for senior O-5s and O-6s.

As events began to unfold at the World Trade Center, reports filtered into Hathaway's sequestered personnel board. He called over to the Pentagon, where his Command Center staff was observing the scene in New York City on large screen televisions and keeping the CNO apprised of the situation. It wasn't long before someone again interrupted Hathaway's meeting and told him that smoke was coming from the Pentagon. He moved to a window on the side of Headquarters that faced the Pentagon to see a column of smoke coming from the building. At the same time, he heard a CNN report that a plane had flown into the structure. He knew from looking at the scene that the impact had to have been near the Navy population in the building. He immediately tried to call over, but got dead phone lines.

Hathaway then called his other office, at the Naval Criminal Investigative Service headquarters at the Washington Navy Yard, which was associated with his AT/FP duties. His staff there confirmed for him that an airliner had flown into the Pentagon and that it appeared that the Navy Command Center had been destroyed. He was told that the other Navy N3/N5 spaces had not been hit as badly and that the rest of the staff had been evacuated.

Hathaway had a small office in the Navy Command Center and another on the fourth deck in the N3/N5 spaces, both filled with the small mementoes, awards, plaques, command coins, and photographs that an officer collects over a long and successful career. He never made it back to either space. Eventually, all of those decks were completely destroyed. All of the office spaces, computer systems, historical files, all were eventually consumed in the fire and the flooding that ensued from the effort to contain and extinguish the fire.

These losses might have been sustained were it not for the human toll that came with them. When terrorists flew American Airlines Flight 77 into the building, there were about fifty people in the Navy Command Center. Within a matter of seconds, forty-two were killed by a combination of blunt force and immolation. Twenty-seven worked directly for Admiral Hathaway. The jetliner impacted almost directly into the Navy Command Center spaces, inundating them with burning jet fuel. The sprinkler system switched on, even as it was overwhelmed.

Because they had just moved into these spaces, many of the survivors were uncertain of the best route on which to make their escape, not just to fire exits, but away from office spaces where everything familiar had been destroyed or turned on its ear. Amongst the many ironies of that morning, one of the few survivors was by far the oldest worker on Hathaway's staff, a budget analyst in his late 60s. For that employee, a wall nearby simply disappeared, and the man walked out into a courtyard and safety.

For awhile, the Coast Guard thought it had lost Admiral

Hathaway himself. So secretive was the nature of Coast Guard personnel board membership that when the Commandant ordered an accounting of Coast Guard personnel at the Pentagon, the first report he received was that all Coast Guard personnel had been accounted for with the exception of Admiral Hathaway. For awhile no one could find him, even as he was sitting only one floor below the Commandant's office.

For Hathaway, the morning had accelerated in milliseconds from a general attack on the country to a direct attack on his personal command. Even with his new responsibilities for the Navy's anti-terrorism and force protection actions, and with the October 2000 attack on the U.S.S. *Cole* in Yemen fresh in everyone's mind, nothing on Hathaway's threat board prior to 9/11 pointed to a coordinated terrorist attack on the American mainland. Perhaps because of the attack on the *Cole*, most of the Navy's force protection focus prior to 9/11 had been on threats to Navy forces overseas.

Yet the attack fell upon him in innumerable and unpredictable ways. Hathaway's N34 duties placed him in charge of the personal security detachments that protected senior Navy leadership, and these teams now moved into action. But no existing plans had anticipated such an in-country contingency. Secure locations away from the Pentagon were hurriedly improvised (plans Hathaway and his staff have rewritten completely in the months since, to focus on in-country threats as they now exist in a post-9/11 world). Senior Navy leadership was removed from the Pentagon, so when Hathaway drove from Coast Guard Headquarters to his office at the Naval Criminal Investigative Service at the Navy Yard (having to shout his way past the gate where no one recognized the new Coast Guard admiral in charge of Navy anti-terrorism), he was only slightly surprised to find the Secretary of the Navy himself in his office. It was the first time Hathaway had met him. The Secretary apologized for commandeering the Admiral's office, to which Hathaway responded: "Sir, it's your office now." After he had introduced himself and explained what he did, the Secretary remarked: "It's a big job you've got ahead of you. Now tell me, Jeff, how did this all happen?"

For the moment, Hathaway had no answers. But he was

determined to get them, and do everything he could to make sure it never happened again. Over the next few hours, he set up secure video teleconference facilities for the Secretary and the CNO, while working to place the Navy in a force protection mode around the globe. With no attack of this kind anticipated on the Pentagon, no common mustering areas had been devised. Much time was spent in accounting for the now-displaced N3/N5 staff, and relocating the OPNAV staff to the Navy Annex that sits above the Pentagon near Arlington National Cemetery. Without a desk, computer, or phone, with many of their historical files gone, this now became Hathaway's temporary home.

Armed with little more than three cell phones, Hathaway and his staff proceeded to conduct Navy business. The N3/N5 staff was now gearing up for a global war on terrorism, even as it dealt with the loss of forty-two of its own. The entire watch section of the Navy Command Center had been wiped out, along with the Command Center itself. A temporary center was stood up within the Marine Corps Command Center, which was in itself an older, smaller version of the destroyed Navy center. The Marine staff moved to one side and allowed the Navy staff to hunker down in their spaces, for what turned out to be months.

As the charred areas cooled, and FBI teams combed through the wreckage, they reported finding massive amounts of classified materials. Dressed in a chem-bio suit, Hathaway returned to the Pentagon, to what was now a federal crime scene, three days after the attacks. As part of a team searching through this classified data, he recognized nothing of the Navy Command Center. What classified material Hathaway did find came from safes that had been melted or blown open. All but bits and pieces of the Command Center's files had been burned.

The next few months became a nightmare of activity. Between the existing and new problems of Navy force protection around the world, and standing up a new command center, there was Hathaway's primary task of seeing to the families of the few injured and the many dead. "I have learned an intimate knowledge of Arlington National Cemetery, leading too many funeral caissons from the chapel to the burial sites. Virtually everyone killed who was buried at Arlington was buried in a plot that overlooks the wedge of the Pentagon that was destroyed. Every time we marched to a funeral site, we were looking into the open gaping hole in the Pentagon."

The blows kept coming, almost more than could reasonably be expected that one man could bear. A much beloved and recently retired Navy O-6, a P-3 pilot who had served both admirals Utley and Carmichael as their deputy, and who had been hired as a civilian in the Navy's counter-drug office, was never found. His became one of the last of the Pentagon funerals, as the family waited in vain for some trace of him to be uncovered.

One of the toughest moments for Hathaway was about a week after the attack, when a staffer told him that one of the victim's sons was outside on the sidewalk of the Navy Annex and wanted to see him. A nineteen-year-old university student just like Hathaway's own son, he told the admiral that his mother had sent him to find his dad, and not to come home until he did. Did the admiral think that any of his father might ever be found? The Admiral took the young man to a bluff overlooking the Pentagon, and tried to explain to him that his dad was gone. What was left of him was down there in the rubble. They talked for a long time. His mother had sent him on a quest, and the boy needed to know what to tell her.

It was an experience Hathaway would repeat in one form or another with many of the families who had suffered instantaneous losses on 9/11. Perhaps no flag officer in the history of the U.S. Coast Guard has had to cope with such an overwhelming combination of responsibilities and losses. When he arrived at the Pentagon in the summer of 2001, he had looked forward to a couple of almost routine years, dealing with issues of staffing and policy. He had actively joined the Navy's pre-9/11 reorganization within the Pentagon, in the hopes that his tenure might be more interesting. Then came 9/11. In its wake, he found himself in many ways uniquely prepared to deal with such a crisis, coming as he did from a Service that was essentially at war each and every day. The Coast Guard existed in a perpetual sense of emergency: be it an oil spill in

Narragansett Bay, a fishing vessel in distress in the Bering Sea, or a drug bust in the Windward Passage. While others argued about what should be done in a particular situation, he took pride that they were usually interrupted by the Coast Guard already off doing it. Now, like the Coast Guard in the drug war, the Pentagon found itself in a war with a shadow enemy that attacked soft civilian targets, fought indecisive battles that led to inconclusive victories, and left the field with no end to the conflict in sight.

Nor was Hathaway unmindful of the fact that he was new on board, and that everyone was looking to the new admiral to see how he handled the situation. In the end, as a commanding officer, Hathaway had little choice but to get back to work immediately and to stay on the job come what may. He felt the rank and file searching for someone to put the attacks into some kind of perspective, to say definitively what course of action was needed and show how it would be done. Hathaway saw fear in people's faces, fear he felt himself. Using his experience as a Coast Guard officer, he deliberately tried to be a figure in whom others could take comfort and derive the will to move on.

As for himself, the high pace of activity, his training and experience, all helped keep many emotions at bay. But Hathaway would find them creeping in at odd moments. Twenty-four hour watches that included a flag officer were established early in the emergency, and the Admiral found himself pulling his share of mid-watches. He would often find himself in a quiet Navy Annex office, in the middle of another slow and demoralizing night, at a time when it was tough to reach out to anyone else. During such times, Hathaway pondered his losses and sought to divert his anger into the plans necessary to prevent such a catastrophe in the future. The next war was now here, and a Coast Guard admiral found himself right in the middle of it.

Chapter Seven

9/11 and the Headquarters Directorates of Marine Safety and Environmental Protection, and Operations: Paul Pluta, Terry Cross and the "M" and "O" dichotomy

Port security concerns were nothing new to Rear Admiral Paul J. Pluta, even before 9/11. He had been Captain of the Port at Marine Safety Office (MSO) Wilmington in North Carolina during the major load-out operations for Operation *Desert Shield/Desert Storm*. As Commanding Officer of the Coast Guard Training Center in Yorktown, Virginia, Pluta oversaw the creation of the schools of port security and intelligence. Pluta went from Yorktown to become Chief of Staff in the Ninth District, which then hosted all three Reserve Port Security Units, each of which deployed to Saudi Arabia and Bahrain during *Desert Shield/Storm*. A hybrid of these units was deployed to Haiti during the migration crises of the mid-1990s. As a flag officer, Pluta was Director of Intelligence and Security for the Department of Transportation, before becoming Commander of the Eighth Coast Guard District in New Orleans.

His intel background, with its daily necessity of reading briefs and threat assessments, learning all the strange occurrences that never found their way into the newspapers, had given the admiral a keen appreciation for the global range of threats arrayed against the United States. His experience had also led him to the conclusion that the Coast Guard's Captains of the Port needed a better appreciation for the risks they were under, along with the legal responsibilities they assumed for the protection of American ports. One of his first moves as district commander in New Orleans was to ask for a counter-terrorism plan from each of his Captains of the Port.

In a pre-*Cole* world, to say nothing of a pre-9/11 world, resources to protect the ports were few. At least, thought Pluta, contingency plans could be in place before an event happened. Such plans, revealing as they inevitably would the weaknesses in port security, would lead the captains to seek wider and beneficial partnerships with both government and private interests. In particular, this might lead them into contact with more intelligence sources, who were not normally part of the port safety universe. It would also serve to offset some of the periodic waffling of the Coast Guard on the role of its Reserve, which found itself caught between the need to prove its value to the Service through daily augmentation, and the national necessity for defense readiness through training for mobilization.

Amongst the many briefings Pluta received after arriving at Coast Guard Headquarters, as the new Assistant Commandant for Marine Safety and Environmental Protection, was one on port security. Nothing in that briefing—the loss of capability, the fragmentation of expertise, the loss of organizational critical mass—surprised him. Neither did the proposed fix—the creation of an office within his directorate for port security, to begin to address ten years of neglect—which he immediately approved. That was in June, 2001. Unbeknownst to all, time was short. But when 9/11 happened, Pluta's staff was ready to hit the ground running with now essential recommendations to recapture the Service's lost port security apparatus and expertise.

On 9/11 itself, Pluta was in Beijing, China, as the vice chair of a conference of nearly twenty Asian and Pacific nations gathered to talk about port safety issues. Along with several Canadian counterparts, who also suddenly found themselves with a basket of new responsibilities, Pluta was forced to wait for four days until a flight back to North America could be found. A senior officer with maritime security responsibilities, the Admiral was distraught while forced to watch events unfolding at home with no ability to get home and contribute. He noted with some reassurance the expressions of solidarity that appeared in the Chinese press. Pluta returned to Washington to find his world of work turned upside down. All of those things that had been so important on 9/10, whether it was passenger vessel safety, the battle against invasive species, or Great Lakes pilotage, all suddenly took a back seat to the specter of terror attacks on the ports.

Among the issues on Pluta's desk was the reintroduction of weaponry into the MSOs. Over the course of many years, it had come to be seen as unnecessarily intimidating for Coast Guard personnel serving at MSOs to carry weapons during routine marine safety checks. With Streamlining, and the ensuing reductions in Coast Guard budgets, a weapons locker at an MSO came to be seen as an expensive irrelevance, especially when, unlike the law enforcement side of the Service, an MSOs day-to-day business did not involve the use of or necessity for weapons. "Because there was no apparent need for them, because there was a lot of overhead involved in maintaining the weapons, having a secure place to store them, with alarms, having Gunners Mates at the units that were needed more elsewhere, and having to run people through periodic qualifications so they could even use the weapons, all of those overhead issues were what drove weapons out of the MSOs."

Given these budget realities, which led invariably to cultural consequences for the Service, Coast Guard personnel who saw themselves as either "M" or "O" people in the three decades prior to 9/11could be forgiven for thinking that they belonged to two separate Services. In Pluta's view, much of this cultural rupture had been repaired prior to 9/11, at several levels of the organization, to the benefit of unifying the Service's vision of itself. The post-9/11 focus on a common goal of port security had the effect of cementing these organizational relationships even further. "When you need people who know about the safe operation of ships, who know what is normal or abnormal on board a ship—whether it's the people part or the documentation part or the cargo part—and you need to put that all together to do good port security, it obviously draws people together as part of the same team, and that has a way of homogenizing the Coast Guard in a way that is very positive."

It was in part this impulse, during the creation of the domestic counterpart to the PSUs, the Maritime Security Teams (MSTs) which led to Vice Admiral Thomas Collins suggestion that these units reflect a safety component as well. They became, following the multi-mission nature of the rest of the Service, Maritime *Safety* and Security Teams (MSSTs), and comprised roughly seventy active duty and forty Reserve personnel. Such tactical decisions were commonplace in the Coast Guard, which has long feared that single mission units were a trap and would be lost as soon as budgets invariably tightened.

Those budgets would constrict in an environment now referred to as the "new normalcy." At headquarters, this meant gathering all of the disparate corners of the Coast Guard to produce a baseline of missions and personnel, platforms, spares and support structures that would represent the new requirements for maritime homeland security. Once Pluta's working group had 'scrubbed' these numbers, they amounted to 4,000 new active duty personnel and nearly 10,000 new reservists. Before 9/11, such increases would have been unthinkable. After 9/11, the Coast Guard received two supplemental budget increases within the fiscal year 2002 budget; for more people, assets, and money. The Service was asked to provide additional requests for the 2003 budget as well. In Pluta's view, undoubtedly shared by all the top leadership, none of these supplemental budget requests would have stood much of a chance were it not for 9/11.

The new resources would be put to immediate use in securing cruise ship terminals and container ports, the latter involving a heavy partnership with the Customs Service. The container issue was at the forefront of Pluta's many new concerns. Working with the International Maritime Organization (IMO), he was pushing for standards for tracking and tamper-proofing containers, from their point of origin to their destinations in the U.S. This problem, in particular, involved American ports in a global container security regime, because without foreign port security, domestic port security became all the more difficult. Pluta envisioned Coast Guard port security liaison officers in the ports of the United States' major trading partners, similar to the current Civil Aviation Security Liaison Officers (CASLO) that the Federal Aviation Administration places at foreign airports to ensure that International Civil Aviation Organization security standards are followed, to perform a similar function for IMO port security standards.

Pluta encapsulated the new reality of his work in a memo to Chief of Staff Vice Admiral Timothy Josiah in early October. On his notepaper, Pluta had inserted the word "Security" into his title, so that it now read: "Assistant Commandant for Marine Safety, *Security*, and Environmental Protection." The word had been a part of the title in years past, but had drifted into disuse, as with actual port security and all that went with it—weapons and their qualifications, and so forth—and was eventually dropped.

Pluta's insertion of the word in his note was a recognition that security was back to stay as a primary mission of the Coast Guard, and his handwritten modification was eventually approved as the new name for his Directorate. In part, this was nothing more than a Coast Guard officer's natural flexibility. Unlike another Service where a war-fighting model might not change for years, if not decades, the Coast Guard universe was far more complex. Being an armed force as well as law enforcement and a regulatory agency, the Coast Guard's battle-space was constantly shifting along a use of force continuum that started with officer presence and worked its way up to deadly force. It was axiomatic that the Service a Coast Guard officer retired from would be a far cry from the one he or she joined. The mix of missions shifted according to budget cycles and priorities, and these were little more than a recognition of what the American people wanted from their Coast Guard at any given moment.

> If I were inflexible I would have found something else to do for a living. That's what keeps us fresh, especially in our program of marine safety and security and environmental protection. You have to be at the forefront of what's going on in the community that you serve. You know that the commercial industry isn't standing still. They're not going to be making money if they don't press the bounds of technology and operations and environment. So, as they learn, we have to keep up with them in

order to deal with all of the safety and security and environmental ramifications of what they're doing.

At the same time, 9/11 caused the Service to look more and more internationally, in search of solutions to all of the security problems associated with global trade. This invariably will lead to more Coast Guard interaction with America's maritime trading partners, in order to tighten maritime security around the world.

*   *   *

Rear Admiral Terry M. Cross had served as Assistant Commandant for Operations for almost two years prior to 9/11, and was in his office that morning when the Command Center called with the news of a "light plane" flying into the World Trade Center. A Coast Guard aviator, he looked out his window at the crystal clear morning, and knew instinctively that there was a problem. He went down one floor to the office of Vice Commandant Tom Collins and they were discussing ways in which the Coast Guard might assist the situation in New York, remarking as he did so that airplanes just do not fly into buildings on clear, calm days, when the second aircraft hit. One of the other officers in the room heard Cross say: "This is a massive failure of intelligence."

An Incident Management Team was stood up immediately, while Cross joined the rest of the Service's senior leadership in plotting a way ahead. Cross had his own personal concerns, with his wife working at the Department of State, where a car bomb explosion was earlier reported. In fact, the Commandant's first request for a Reserve mobilization included mention of this attack, which later proved to be only a false rumor.

For Cross, the type of attack might have been surprising, but such an attack had been part of Coast Guard strategic thinking for some time. As late as January of 2001, the Service had signed an agreement with Defense, State, and INS, to improve what had come to be called "Maritime Domain Awareness" (MDA), especially with regard to intelligence-gathering on weapons of mass destruction

(WMD) that might be smuggled into a port on board ships or in a container. In the immediate aftermath of 9/11, as the Coast Guard executed a massive shift from offshore work to inshore port security, Cross's concerns shifted as well. First, operational commanders required more resources to maintain the left turn toward port security, while not burning out their aviation and surface platforms and personnel. At the same time as they were feeling a personnel crunch, the Coast Guard sent personnel to the new Office of Homeland Security to stand up a situation room for the incoming director, Governor Tom Ridge of Pennsylvania. Similar assistance was given to the new Transportation Security Agency (TSA).

The goals of the Coast Guard's initial response were six-fold: first, to control the movement of all traffic in the ports and along the waterways while, second, focusing on high risk vessels such as tankers carrying gas, oil, chemicals, and propane, and, third, passenger-carrying cruise ships and ferries. Fourth, to identify and develop security schemes for critical port and waterways infrastructures from the Golden Gate Bridge to the Statue of Liberty. Fifth, to assist the crime scene teams in New York harbor, and, lastly, to develop schemes to identify those vessels, crew members, and cargoes that raised the concern of the Service, and inspect them accordingly.

This new normalcy of heightened security required a long-term strategy for the Service and its relationships throughout the maritime world. This new strategy meant seven-day work weeks for the staffs of both Rear Admiral Pluta and Rear Admiral Cross. Through a number of iterations, the framework for a long-term policy evolved rapidly. It required the "M" and "O" staffs to define all of the elements involved in the new normalcy in terms of mission requirements and associated operational activities, and at what level they would have to be funded. Counter-drug patrols had dropped to twenty-five per cent of their pre-9/11 levels, while fisheries enforcement dropped to nearly zero per cent. Neither situation was in the long-term national interest. Any 'new normalcy' funding for the Coast Guard would have to get these patrols back to their pre-9/11 levels, in addition to funding all of the new port and waterways security initiatives.

Despite the awareness of the senior Coast Guard leadership to potential threats from terrorism within the ports, the Service had been able to gather only $14 million in the initial fiscal year 2002 budget for maritime homeland security. After 9/11, this figure was increased twice, first by $195 million, then by another $126 million. By mid-fall, 2001, the Coast Guard was apportioning $335 million in funding in nine distinct maritime homeland security categories. These included intelligence ($23 million), connectivity between units ($18 million), port vulnerability analysis ($31 million), increased harbor patrols ($85 million), boardings and escort operations ($7 million), creating and deployment of new Maritime Safety and Security Teams ($22 million), increased weapons, weapons qualifications, and ammunition ($6 million), force protection ($17 million), and support and logistics ($125 million). By definition, these kinds of increases, centered as they were on port security operations, served to bring the marine safety forces of Pluta and Cross' force of operators even closer together, a process which had been ongoing in the years prior to 9/11.

The Coast Guard made rapid progress on the intelligence front, as well. In Cross' first years as Assistant Commandant of Operations, he learned, after the fact, of information on threats to American ports that was held in several other federal agencies and never passed to the Coast Guard. In some cases the intelligence was passed to the Navy, which had no authority or responsibility for port security. As a result of 9/11 and the imperative for intelligence sharing, the Coast Guard finally became a formal member of the U.S. intelligence community by Presidential decree on 28 December 2001. Intelligence gathering in the Coast Guard was located within "O," with a clear recognition of "M" as one of the primary customers for the products of intelligence. Plans call for intelligence fusion centers in both Area Commands, along with human intelligence gathering in the field.

> We needed to enhance port security. Most of the Captains of the Port are marine safety officers. But in order to enhance port security you need

the kinds of skills that operators have: boat drivers, law enforcement, port security, and intelligence.   These are all programs managed by Operations.   So we had to come together [with Marine Safety] even more than we had previously in Headquarters, to provide planning, programming, and budgeting.   In the field it became absolutely critical to have a unity of effort in our ports.   Even before 9/11, we were committed to reaching out to other agencies in the ports, so after 9/11 we became the leader/coordinator of efforts.   There aren't enough of us to do all the things that need to be done.   We weren't at the nadir, but we were pretty low [before 9/11] in terms of people in the active duty Coast Guard, about 36,000.

After 9/11, a more robust intelligence capability assumed such importance to the Service that Admiral Collins created a new and distinct position, the Assistant Commandant for Intelligence (G-C2). This two-level elevation within the bureaucracy, along with its designation as "C2," was an indication that the service was moving toward a more security-conscious, Navy-style battle staff, with C2 reflected in the Navy's intelligence staff designation of N2.

For Admiral Cross, the theme that would ultimately unify the operators with the marine safety people (logically in a combined "C3" operations designation) was that they were both lifesavers. Where differences existed, they did so in the approaches of the two sides of the Service: marine safety officers focused on prevention of loss at sea, while operators trained to respond once a disaster had already taken place. But even within this definition there are anomalies, such as the National Strike Force, a top-of-the-line "M" program with a clear mission of response. And even as a career operator, as Commander of the Seventeenth Coast Guard District based in Juneau, Cross had stressed prevention. Prevention saved lives, saved dollars, and reduced the Admiral's sleepless nights worrying about aviators and boat crews out in eighty knot winds and forty foot seas.

Such prevention, or rather the loss of prevention capabilities, was amongst Cross' primary operational concerns in the initial stages of the 9/11 response. It is true that Coast Guard search and rescue (SAR) operations fell off after 9/11 as a result of the shift to port security, but the bare fact is misleading. The Coast Guard responded to all SAR cases; but, with the ports and waterways locked down and Americans fearful of traveling, the number of SAR cases dropped as a result of fewer people being on the water. Cross was more concerned with the shift of the buoy tenders from Aids to Navigation (AtoN) work to providing a Coast Guard presence in major ports. If AtoN work was ignored for any appreciable length of time, then any unattended navigable aid discrepancy could lead to navigational error on the part of a tanker and perhaps result in a disastrous oil spill. Balancing the risk of maritime disaster versus maritime terrorism became a part of everyday thinking for Coast Guard planners.

The new normalcy, represented by the first of the three distinct maritime security conditions developed by the Service's leadership, MARSEC ONE, attempted to address these mission requirements by leaving the cutters offshore. MARSEC ONE required primarily an enhanced intelligence capability on the part of the Coast Guard, along with more small boat presence in the ports—of necessity, more small boats and boat crews—as well as increased air presence in the form of helicopter patrols. This first condition operated in large measure without recourse to the cutter force, freeing the fleet for the operations that had been so constrained after 9/11: counter-drug and fisheries patrols and AtoN work.

The Service's leadership was mindful as well of the links between drugs, migrants, organized crime, and terrorist organizations. All of these Coast Guard missions, along with fisheries patrols, were suddenly recast in terms of security: national, economic, and border/homeland. Getting the fleet back offshore fed into this integrated strategy for overall maritime homeland security, because it allowed the Coast Guard the chance to intercept problems before they penetrated the American maritime frontier. Admiral Cross saw this pattern as one that had developed over half a century.

> Fifty years ago, the Coast Guard was a response and consequence management organization. If people got into trouble, or there was a law enforcement problem, they called us, and we'd respond. On those occa-

sions where we failed, we'd try to clean up the mess—but we tried not to fail very often. Maybe twenty to thirty years ago we recognized the idea of prevention. We had the commercial vessel safety act, recreational boating safety act, so we became a prevention, response, and conse-quence management organization. In the past three or four years, we recognized that a piece was missing and that piece was awareness. We were trying diligently to become an awareness, prevention, response, consequence management organization, but there was no momentum. 9/11 provided that momentum.

Chapter Eight

9/11 and the Atlantic Area District Commanders: Admirals Carmichael (D7), Casto (D8), and Hull (D9)

1. 9/11 in District Seven

Around the conference table in the office of the Commander of the Seventh Coast Guard District Office in Miami, Florida, the morning brief on 9/11 got underway, as it did each day, at 0830. In this most operational of Coast Guard Districts, the gathering of senior staff was invariably filled with news of the latest law enforcement or search and rescue mission. For the district commander, Rear Admiral James S. Carmichael, there was never any shortage of current material to talk over. Nevertheless, he and his staff always took the chance to look over the horizon and discuss the things they felt the district would confront in the days and months ahead.

As in most offices around the Service, it was the impact of the second aircraft into the World Trade Center that immediately redirected the everyday business of law enforcement and search and rescue into port security. The attack on the Pentagon soon after had a direct impact on Admiral Carmichael, who had once occupied the same Pentagon billet filled on 9/11 by Jeffrey Hathaway. At first, his concern was mitigated when he heard the spot where the aircraft had crashed, because it was not near the area where he had worked. Only later did he learn that the Navy Command Center had moved since his time there, and that several dear friends and colleagues were killed.

Admiral Carmichael and his staff began to catalog the high risk facilities that existed within the Seventh District, and the list was not a short one. Defense facilities alone constituted a major waterfront security challenge, and included a naval base in Mayport, Florida, a nuclear strike "boomer" submarine base at Kings Bay, Georgia, as well as the three strategic military out-load ports in Charleston, Savannah, and Jacksonville. Carmichael wanted to ensure that his forces were positioned to provide escorts and waterside security as requested by these facilities.

Carmichael and his staff then turned their attention to high-capacity passenger vessels such as cruise ships. To head off any *Cole*-like attack, Carmichael offered them protection while they were at the docks. This was particularly important in Miami, where small boat traffic passed by large cruise ship terminals each day. Such dockside protection evolved quickly into armed boardings at the sea buoy—accompanying the pilot to the bridge to make certain that these vessels were not used as weapons as the airliners of 9/11 had been. Unlike the long transits from the sea buoy in, for example, New Orleans, those to the southern Florida cruise ship terminals were very short. In Miami, Port Everglades, Port Canaveral, or those in the Greater Antilles Section (GANTSEC) ports of San Juan, Puerto Rico, and St. Thomas, U.S. Virgin Islands, the sea buoy transits were on the order of two miles or less. This led to the use of Seventh District vessels to escort cruise ships from the sea buoy to the passenger terminals, where a security zone was established around the cruise ship. When more than one cruise ship was in port, the area surrounding the terminals was closed to small boat traffic.

Admiral Carmichael delegated authority for warning shots and disabling fire to his Group Commanders, and allowed them to further delegate this authority down to the coxswains. "We had never before in history delegated that kind of authority that far down the line," said the Admiral. "That put a lot of responsibility on a very young person." As for the sea buoy boardings, they too were something of a new experience for the district.

Over the course of time, we had lost the capability in the marine safety program, for armed boardings. That's where we found ourselves on the 11th of September. Any kinds of armed boardings that had to be done for any particular reason were done by our groups and stations. Very rarely had the MSOs called the groups and stations out to do those kinds of boardings because [of what] we were involved in on the marine safety side of the house. The Ports and Waterways Safety Act, as well as some of the Caribbean cargo safety codes that we enforce, these were invariably done at the pier by marine safety officers, as opposed to being done at sea.

Immediately after 9/11, Admiral Carmichael knew he needed both the expertise of his MSOs, combined with the security of armed boarding parties. In the Seventh District, that meant that the commanding officers at the Groups and MSOs had to sit down together and put together a team that possessed both marine safety and boarding skills. Where MSOs and Groups were not co-located, Carmichael and his staff envisioned a kind of virtual joint operations centers, where the skills of the MSOs could be made operational. Captain James W. Stark, D7's Chief of Operations and a former Group Commander, had observed in the past how mission effectiveness depended upon personal relationships built up between the Captains of the Port, the MSOs, and the Group Commanders. "Since 9/11, we've seen that it's really a mandatory lash-up. The way the Coast Guard has evolved, the Marine Safety Office doesn't have the infrastructure or the support to stand up the operational capability that it needs to do this mission."

Paul D. Kirkpatrick, a Reserve captain who as a civilian worked as an aerospace engineer for the National Aeronautics and Space Administration (NASA) at Kennedy Space Center, arrived at Seventh District offices in Miami on the morning of 9/12, intent on finishing out a few drills as the temporary head of Marine Safety. Instead, he did not leave until the end of January, 2002. Like Captain Richard Sullivan at Eighth District headquarters in New Orleans, he found himself overseeing maritime homeland security missions. Within a week, these included an expansion of the Coast Guard's Cold War-era Special Interest Vessel (SIV) Program, the search for inbound ships that might pose a danger to the American maritime frontier. The SIV program was based on the nationality of the ship, not the crew, and any vessel coming from a particular country would then be "of interest" to the Coast Guard. A new "High Interest Vessel" (HIV) program began to place individual crew members under a security microscope.

Such programs were a natural outgrowth to long-standing marine safety concerns. If weapons lockers were not to make a dramatic reappearance at the MSOs—and both Admiral Carmichael and Captain Kirkpatrick had their reservations about such a shift—then the interaction between the Coast Guard and the civilian and corporate worlds on a regulatory and inspection level depended on more and better intelligence, not necessarily more and better weaponry. For the Admiral, this was nothing more than elementary mathematics.

> We are a wide open country. You're never going to arm enough people, or have enough boats and security to close down the borders. That would be totally dysfunctional, in any case. It would stop commerce. In my view, there's no need for all marine safety inspectors to be armed. Every Coast Guardsman is a humanitarian, a public servant committed to enforcing maritime laws, to rescue people who are in distress, and to ensure the safety and security of the marine transportation system, to protect the viability of this country's economy. It's an abiding concern for the United States citizen, whether that's doing a boating safety inspection, or boarding a suspicious vessel to take drugs out of the system, responding to someone lost at sea, or checking the safety of a commercial vessel so that it doesn't spill oil into American waters and cause environmental or economic havoc on the beaches.

As for security measures the Coast Guard was able to bring into play after 9/11, how long they would last, in a Service historically short of personnel, platforms, and operating budgets, remained to be seen. For Admiral Carmichael, who gave a speech on December 7, 2001, that drew parallels between Pearl Harbor and 9/11, the ever-present travails of the Coast Guard budget left open too much room for people to let port security languish. In the Seventh, for example, only one billet was devoted to planning and security studies. In a conference of Atlantic Area commanders held a month prior to 9/11, port security had not been in the top six areas of concern discussed.

To counteract such complacency, Admiral Carmichael believed it was up to the Service to continually explain the vulnerabilities of the maritime system, and constantly reinforce the partnerships within the ports that acted as a multiplier for Admiral Loy's vision of Maritime Domain Awareness. After 9/11, the largest immediate force multiplier for the Seventh District was the Coast Guard Auxiliary, the largest number of auxiliarists of any Coast Guard District. At one point after 9/11, 85,000 out of 150,000 auxiliary volunteer hours were donated by Carmichael's auxiliary in the Seventh.

In Jacksonville, Captain of the Port Captain Michael Rosecrans and Group Commander Commander Mark Wilbert became the center of gravity for the development of a risk-based assessment model that decided which threats deserved what level of scarce resources. Their joint operations center tested the model on an array of scenarios. When these scenarios were run against it, they were able to develop resource levels targeted to those specific threats. A training team exported this model to other Captains of the Port, to attempt to bring a level of consistency to threat assessments within a constricted budget universe. In St. Petersburg, Captains Allen L. Thompson and Fred M. Rosa developed a four-layered continuum of protection construct that ranged from protection to defense, a construct that also helped as a model for resource allocation. Like the Jacksonville model, the St. Petersburg model was exported to the area command as well.

If 9/11 found officers of equivalent rank from the operational and marine safety sides of the Service thrown together for a joint operation—with the jockeying for position in the chain of command that that could entail—the Seventh District dealt with this by recourse to the incident command structure. The lead officer was designated ahead of time, depending on the type of emergency encountered. In a hurricane, for example, the group commander would become the lead officer. In the event of a spill of major significance, that role would be filled by the Captain of the Port.

On 9/11, the port-centric nature of the crisis led to the Captains of the Port becoming the lead officers. This was somewhat anomalous for the Seventh District, where the majority of Coast Guard units were organized to carry out search and rescue and discretionary law enforcement involving fisheries violators or smugglers coming across from the Bahamas.

Captains of the Port were empowered under Section 22 of the Code of Federal Regulations (CFR) to establish security levels for cruise ship terminals. Even before 9/11, the State of Florida had passed a seaport security bill, directed primarily toward pilferage and smuggling, but terrorism had been incorporated into the study that led to the legislation. Security inspections of the ports mandated by the legislation had just begun at the time of 9/11. On 9/11, the Captains of the Port went immediately to the highest security level with the port authorities and the lines operating the passenger vessels. Admiral Carmichael saw it as a dramatic operational, as well as psychological, shift for the district.

> We have always felt that the threats coming at us down here require us to be in a forward presence mode offshore. 9/11 brought all of our resources circling the wagons inside the ports. It was clear that if we stayed in that posture for very long, an awful lot of threats would be coming from migrants and drugs and to a certain extent from fisheries violations as well.

The district was helped in the first few weeks after 9/11 by a dramatic drop in calls for search and rescue, as well as an absence of reports of migrant landings. The Admiral ascribed the latter situation to a possible hunkering down effect after 9/11, where criminals decided not to challenge American power at a time when the U.S. public would not stand for any such challenges. Beyond this, the Admiral saw all of these elements of Coast Guard operations—drug and migrant interdiction, fisheries and other economic patrols—as parts of a holistic approach to American maritime border security. The difference between Coast Guard operations before and after 9/11 was the nature of the maritime border itself, which the Coast Guard had for many years pushed further offshore and into the

Caribbean.

After 9/11, Carmichael viewed the regular flag conferences as a way to initiate discussions over best practices and new ways of solving old and familiar problems, in a Service that was continuously operational and therefore chronically short of time for strategic thinking. Once the immediate pressures of port security operations after 9/11 began to ease, Commander Michael J. Scully of the Seventh District's law enforcement planning staff—who had been forced to scrap pre-9/11 force lay-down packages and budgets—began to work through the numbers of hours that were left on the district's resources. This was essential if the district was to begin to return to pre-9/11 missions and operational levels.

For Captain Stark, such work enabled the Seventh to lead the way for the rest of the Coast Guard, providing an example of how the Service could return to pre-9/11 missions while at the same time meeting the new responsibilities for homeland security. Coast Guard planners are traditionally wary of promises of large budget increases—rumors that were rampant after 9/11—so it came as little surprise when the fleet was eventually asked to keep to programmed hours for operational platforms, tasked to perform new missions, and cautioned not to burn out personnel.

For Admiral Carmichael, the numbers dictated the operational tempo. "We knew we weren't going to have anything left in the tank if we kept everybody at general quarters forever, especially in the non-descript threat environment that we were looking at." By early November, Carmichael's staff had come up with a new set of operating hours for the district's small boats and patrol boats, one that maintained a homeland security posture that both supported law enforcement and search and rescue, and kept platform and crew hours within some strict budget limits.

In any event, it was not until December of 2001 that many of the cutters that had been recalled from the Caribbean waters began to come back online in their traditional law enforcement roles, with a host of new maritime homeland security responsibilities thrown in the bargain.

2. 9/11 in District Eight

In New Orleans, inside the Hale Boggs Federal Building a few blocks from the French Quarter, Captain Joel R. Whitehead, Chief of Staff of the Coast Guard's massive Eighth District, had just convened, ironically, a joint Coast Guard/Maritime Administration conference on maritime anti-terrorism. The conference started at 0800 New Orleans time, three minutes before the second airliner hit Tower 2 in New York. Alerted by his operations center, Whitehead gave his planned introduction and then politely excused himself. There was a lot of work for the acting district commander to do. In the absence of Rear Admiral Roy J. Casto, who was on leave, Whitehead gathered his chiefs of operations and marine safety, and within an hour they began to redirect all of the district's cutters into position as picket boats guarding the entrance of all of their strategic ports.

The Eighth District was a recent creation, an unwieldy amalgamation of the Eighth and the old Second District centered in St. Louis. The Second District had been another casualty of streamlining, so that the newly enlarged Eighth now encompassed all or parts of twenty-six states, from the headwaters of the Mississippi River and all its tributaries to the Gulf of Mexico, an area comparable to the original Louisiana Purchase. With the reverse logic familiar to members of the Service, Admiral Casto was given forty per cent less staff to do the same job that two admirals and two hundred more staffers had done prior to the consolidation in 1995. As the admiral remarked, most of his staff was still trying to figure out how to get from New Orleans to such unfamiliar outposts as Keokuk, Iowa; Paducah, Kentucky; and Paris Landing, Tennessee; in an attempt to reassure those who believed that the Coast Guard's district consolidation had abandoned the inland rivers. Casto had made this outreach a personal mission when he became district commander in April of 2001.

On 9/11, Captain Whitehead, who had arrived in Louisiana in May of 2001 after a tour as Captain of the Port of Boston, had more than enough 'routine' disaster scenarios from accidental groundings, fires, and oil spills, without the added burden of terrorism. Nearly a fifth of the nation's commerce floated down the Mississippi past or toward six of the country's ten largest ports.

> Some of the things that we wanted to protect immediately were the Louisiana Offshore Oil Port (LOOP), which is twenty-six miles off the coast, and we requested from the Area a medium endurance cutter. We've not been able to talk the Area into having a regular MEC presence in the Gulf. They've all gone over toward the Seventh District and drug patrols in the Caribbean. But we made a request right away for that, and we made a request almost immediately for what now [is now] called an MSST, basically a port security unit that could go to the Houston/Galveston/Port Arthur area because that is a major petroleum/chemical complex.

Whitehead got his MEC. He also hoped that the district would get the Mississippi-based Port Security Unit 308 as well, but this was not to be the case. PSU 308 was already scheduled for overseas deployment, after joint training in North Carolina, and was not pulled from that assignment for domestic security.

When Admiral Casto returned to New Orleans three days after the attacks, the issue was even more basic. He was dumbfounded when news releases appeared implying that the Coast Guard was providing armed security at 361 ports across the nation. He knew that the loss of weapons and weapons qualifications throughout the marine safety community meant that there were few if any guns north of New Orleans. "We had a situation where an activated reservist came on board in Pittsburgh. He happened to be the Chief of Police in Pittsburgh. As the Chief of Police in Pittsburgh he carried a weapon, enforcing the law. We put him in a Coast Guard boat or car [without a weapon] and had him patrol a chemical plant, hoping that if there were a terrorist he could do something about it. But we denied him the ability even to have self-protection by carrying a weapon."

Casto pressed the issue up the chain as hard as he could, but found himself frustrated by the process. Told that it would take months to procure weapons, the Admiral went so far as to spend $513 of his own money to go to a local gun shop and buy an M-9 just to prove that it was neither difficult nor time-consuming to get one of the same weapons his armed Coast Guardsmen were carrying. "I just felt that it was unforgivable to send people out in harm's way, possibly to confront a terrorist, and not have that individual possess the means to protect himself." Part of the post-9/11 supplemental budget given to the Coast Guard went to providing secure weapons facilities for the MSOs. But it would take time to provide both the weapons lockers and the gunner's mates to ensure the necessary weapons qualifications of unit personnel.

The Eighth District's seven group commands and twelve MSOs were deluged with requests for Coast Guard protection of sensitive waterfront facilities. Captain Whitehead recalled that the demand for resources was "a hundred thousand times beyond what we had." Admiral Casto returned to New Orleans to find a pervasive sense that more attacks were coming. "These people used improvised explosive devices—airplanes—and we have a lot of things either based here or that come in and out of here that would make even bigger improvised explosive devices." Without the assets needed to guard everything, Casto leveraged the intelligence coming into the Operations Center in New Orleans to employ the district's resources where they were most needed. Such intelligence, however, was hard to come by, and at times it seemed as if the best information was coming off the television. In the absence of specific threats, one alternative was to triage the highest risk structures no matter who might threaten them. "I think we've done a good job of defining our vulnerabilities," said the Admiral. "The problem is, once you've got all the vulnerabilities laid out, there's no way to shield all of them."

> In the wake of September 11, we started responding to: 'Hey, Coast Guard, this is your job. Come protect my petrochemical plant seven by twenty-four from the river.' 'Hey, Coast Guard, come protect my MARAD prepositioned vessels down here on the river seven by twenty-four.' 'Hey, Coast Guard, come protect the Navy barge that takes people across the river to go to work at the Navy complex.' And we did that, at the outset.

[But] we had to get ourselves out of those things, because it was clear that we couldn't do those kinds of things.

Initial port vulnerability studies revealed nearly 400 high importance or high consequence facilities on or near the water in the Eighth District. Nuclear plants, cruise ships, petrochemical plants, the Huey P. Long Bridge, the old river control structure that kept the Mississippi from turning New Orleans into a bayou, all required intense levels of security. District staff worked with their customers to transition from Coast Guard to private security contractors for these sites, but the often times dubious quality of that private protection led many waterfront customers to prefer the presence of the Coast Guard. Admiral Casto's district held scores of locations that presented the immediate specter of mass catastrophe in the event of terrorist sabotage.

More tactical considerations included the radically different nature of ports in the Eighth District. Where the run in from a sea buoy might be only a few miles in another district, as in the Seventh District, the run from the sea buoy to the dock in Houston was more than fifty miles, to New Orleans more than one hundred, and to Baton Rouge it was 220 miles. A moving security patrol along such a run was extremely difficult to maintain, and Sea Marshals were largely impractical in such a situation. Such geographic anomalies made it very difficult, as well, for Headquarters to issue uniform national maritime security policies, and inadvertently made it seem as if the Coast Guard was offering more protection in ports like San Francisco, for example, than it was for Baton Rouge.

To begin to work through these new problems, Casto stood up a maritime homeland security staff under Chief of Staff Captain Whitehead, headed by Reserve Captain Richard Sullivan. Admiral Casto deliberately placed Captain Sullivan's cadre within the Chief of Staff office, to elevate it above any potential conflicts between the "O" and "M" staffs. This allowed Captain Sullivan, a federal probation officer in civilian life, to begin working long-term strategic security issues with a sense of continuity, one that might have been lacking had they been tasked into the daily operations of Marine Safety or Operations. Outdated port vulnerability studies were dusted off, updated, and upgraded in importance. For each unit in the district, Reserve officers were brought onto active duty as both planning officer and point of contact for Captain Sullivan, in order to cover an area that sprawled north to south from the Canadian to the Mexican borders, and from west to east from Colorado to West Virginia, and contained twelve of the nation's top fourteen ports. Sullivan became Casto's 'go-to guy' for maritime homeland security issues, with the ability to work issues over time, the knowledge to be able to brief the leadership when required, and the authority to take action if necessary.

Assigned for his Reserve duty as Deputy Operations Officer at Eighth District Headquarters in August of 2001, Captain Sullivan was mobilized under Title 10 USC a few days after 9/11, and spent more than a year as head of maritime homeland security for the Eighth District. Besides the giant assemblage of petrochemical plants in the Houston-Galveston area, which processed much of the fuel used by the national economy, many of the ports in Sullivan's new AOR were in fact long stretches of Mississippi River, linear and qualitatively different from what many associate with a 'typical' seacoast port like Seattle, Boston, or Charleston. A Captain of the Port along the Mississippi might have an area of responsibility several hundred miles long and only a few hundred yards wide.

As Captain Sullivan pointed out, one of the largest tonnage ports in the U.S. was Huntington, West Virginia; a vital loading facility for coal and grain as well as specialized and volatile chemicals. The lower stretches of the Mississippi were called 'chemical alley,' because of their concentration of petrochemical industries. The offshore oil patch in the Gulf of Mexico held strategic oil loading facilities. For several weeks, Sullivan found that it was difficult to eat or sleep, and would often get up in the middle of the night to jot down reminders about some facility or critical task. He worried constantly about where the district could most effectively place its bodies, its boats, and its bullets and, even more worrisome, where they could not.

The standards Sullivan set for his team as they built a picture of the critical assets within the Eighth District were high. Such assets had to be critical to the nation in either economic or psychological terms; for example, nuclear plants or bridges on the one hand or the St. Louis arch or waterfront stadiums on the other. Of primary importance were waterways that carried hazardous cargo or passed major population centers, where ships could be blown up, used as a weapon of mass destruction, or driven into another object. He implored his team to "think like a terrorist. Look at your AOR, think like a terrorist, then put up your visible presence and protection at those points that are most vulnerable. You might have 300 targets and can only protect two."

In any event, over 1,300 sites in the Eighth District met Sullivan's criteria and needed protection. After placing these sites on three tiers of importance—by following the model developed by Admiral Nacarra's staff in Boston—Sullivan presented Admiral Casto with a list of 396 critical assets. Grouping many of these sites and "playing zone defense" narrowed the list still further. It was this initial analysis after 9/11 that allowed Casto to place his limited resources where they would have the great effect in hardening potential maritime targets.

It was the kind of staff work both Casto and Whitehead felt the district needed prior to 9/11, but would never have been possible given existing budgets and operational priorities. 9/11 provided the impetus for the new staff, but only for the short term. Unfortunately, the expertise earned by this cadre was threatened by the Service-wide mandate to demobilize, by the end of Fiscal Year 2002, all of the reserves called up after 9/11.

3. 9/11 in District Nine

In Cleveland, at the headquarters of the Commander of the Ninth Coast Guard District—the 'Guardians of the Great Lakes'—Rear Admiral James D. Hull's morning brief was just concluding when the first aircraft hit the north tower of the World Trade Center. Admiral Hull had been stationed in New York as Commanding Officer of the 378' Hamilton Class high-endurance cutter *Dallas* (WHEC 716). For two years, he had had the Trade Center towers over his shoulder, and he was stunned to see them come down so fast.

A thoughtful operations specialist whose career and character were shaped by combat experience in Vietnam and hurricanes in Florida, Admiral Hull reacted with thunderclap speed to the impact of the second aircraft on the World Trade Center. He immediately ordered the more than fifty foreign ships then in the Great Lakes to go to anchor or to stay at their docks. Hull believed that the Coast Guard was the finest first response organization in the country, an impulse derived from what Hull considered the Service's core mission of search and rescue. The primary mission of lifesaver led to the Service's ability to respond quickly and efficiently to all maritime emergencies

Prior to 9/11, vessels entering the Great Lakes were inspected in the St. Lawrence Seaway, but the district commander in his operations center in Cleveland didn't necessarily keep tabs of their subsequent movements. Now, Admiral Hull wanted to know the identity of all foreign vessels maneuvering into the heartland of the country, as well as who was on board, what was the nature of the cargo, and where were they bound. A tightly-knit U.S. and Canadian waterways management forum helped Hull make contact with his counterpart in the Canadian Coast Guard on the other side of the northern border, even as the Canadians were calling Ninth District headquarters and asking how they could help.

No sooner did Admiral Hull begin to secure the northern maritime border, than he was forced to evacuate his own offices. After the first two airliners attacked New York and a third hit the Pentagon, a fourth commercial airliner had abruptly changed its course over Lake Eire and swung back in the direction of Cleveland. Without knowing where this aircraft was headed, the A.J. Celebrezze Federal Building in downtown Cleveland, the home of the Ninth District's offices, was emptied. The streets outside were jammed, so the admiral left his car behind and walked downhill to MSO Cleveland, located on the shore of the lake at the foot of 9th Street. Hull gathered his command center together at the MSO. This arrangement sufficed for personnel, but left something to be desired when it came to back-up communications and computer systems.

Later, when they were able to reoccupy their federal building spaces, Admiral Hull and his staff turned their attention from the bulk cargo carriers on the Great Lakes to the critical infrastructure that surrounded them. The Ninth District consists of over 6,500 miles of shoreline and a thousand miles of international border, so the catalog became an extensive one, including not just obvious items like the massive bulk cargo loading facilities, but nearly twenty waterside nuclear power plants.

Seventy percent of the trade between Canada and the U.S. passed over four bridges and through three tunnels in Hull's District. Much of that trade came in the form of containers which, though driven by truck, depended on essentially maritime infrastructure like bridges and tunnels. There were also a host of refineries and other important economic facilities on the Canadian side that could have an impact on American mobility and commerce, and Hull's staff devoted much attention to these, as well. Hull saw the Coast Guard response in the Ninth District evolve along several related lines, all of which intersected at public confidence.

> We had several roles. The first was to assure the public that we were there doing the best we could, and to get back to normal and get commerce moving as fast as possible. We put our airplanes up and flew them a lot and very visibly. People in Chicago thought we had airplanes up all the time when we were flying them once a day. I made them go down the rivers [to increase their visibility]. What was the first [financial] market back [in operation] in the United States? The Chicago Board of Trade.

The back wall of the Chicago Board of Trade is a bulkhead on the Chicago River. We put people on that river immediately. I never downplayed the effect on the American and Canadian psyche if something were to happen in Chicago. New York is one thing, Los Angeles one thing, but [an attack on] farm country and the breadbasket of America [would] mean that nothing was safe. People thought we were out there probably more than we had the ability to be out there.

Hull divided the district in four different security regions, and appointed 'maritime security commanders' in each of those regions who reported directly to the admiral; two of the commanders were operational officers and two were marine safety officers. The situation was helped in the Sault Ste. Marie area, where a combined Captain of the Port/Group/MSO already had a unified chain of command. In Milwaukee, Detroit, and Buffalo, the Group existed separately from the MSO. Hull viewed 9/11 as a catalyst for more such unified commands as the one at Sault Ste. Marie. "It's the way to do business. We have to grow people as specialists, but when we respond, the people on the outside really don't care. They want the Coast Guard to respond, and we've got to be able to respond without internal barriers."

Like Vice Admiral Riutta in Alameda, who was glad at least that most of the west coast summer cruise ship season was over on 9/11, Admiral Hull could take some comfort that by 9/11 most of the Great Lakes five million pleasure boats were off the water. The concurrent reduction in SAR cases allowed Hull's small boats, which in the summer were operated from some of busiest small boat stations in the Coast Guard, to focus instead on port security patrols. Prior to 9/11, much of the Ninth was staffed for search and rescue and maritime safety. With few extra assets projected for the district, it required resource reallocation to cover the Ninth District's pre-terrorism duties and take on a new collection of maritime homeland security missions.

Where no new resources were forthcoming, Admiral Hull, like Admiral Naccara in Boston, looked to develop much closer relationships with state entities and state governors in his district than had existed prior to 9/11. This allowed the Coast Guard to leverage other potential forces—state National Guard troops, state troopers, and the like—in an event where the Service could not provide blanket coverage to every strategic waterfront infrastructure.

Chapter Nine

9/11 and the Pacific Area District Commanders: Admirals Brown (D13), Utley (D14), and Barrett (D17)

1. 9/11 in District Thirteen

In Seattle, in the nation's third-largest container port, Rear Admiral Errol M. Brown was usually at his desk, as he put it, at "oh dark thirty." He liked the time an early start offered him to quietly contemplate the coming day, and think through all that needed to be accomplished. Three hours behind Eastern Standard Time, he usually caught up on news from Washington by turning on a television hidden behind a chart of the Washington-Oregon coast that hung on a wall in his office. On 9/11, he was preparing for a trip to a fisheries conference in Oregon when he turned on his television in time to see the first plane hit the Trade Center. A few minutes later, after the impact of the second plane, his phone rang. It was the U.S. Navy, asking Brown to send everything he had in order to secure naval facilities in the Pacific Northwest.

In Seattle, where Admiral Brown had become commander of the Coast Guard's Thirteenth District in June of 2000, maritime security priorities before 9/11 became very different in the aftermath of the attacks. The Thirteenth is a large coastal district, with breaking bars and dramatic surf pounding upon high angled cliffs that requires intensive search and rescue work from small boat and air stations. In the large ports of Seattle and Portland, the pre-9/11 emphasis was on expediting commerce and the rapid flow of goods both through the ports and across the border from Canada, and on the continuing post-*Exxon Valdez* priority of environmental protection and the safe movement of oil tankers through the region.

Even so, bits and pieces of the post-9/11 world could be found before the attacks. In the aftermath of the attack on the U.S.S. *Cole*, the Thirteenth District had begun to work much more closely with the Navy to align its force protection procedures. This work followed Admiral Brown's efforts to strengthen his district's emergency preparedness, a program that led to the kinds of intensive Incident Command Structure (ICS) training that was immediately put to the test on 9/11. One of Brown's first tasks was to canvass the outlying small boat stations, to see what assets could be broken out to increase the Coast Guard's presence on the Seattle waterfront. But this could only be accomplished after these stations curtailed operations.

The morning accelerated rapidly, and Brown was soon on a video teleconference (VTC) with Vice Admiral Riutta. With decisions flying, he ordered a junior officer to get a sheet of paper and start to record everything. Brown remembered the lack of organizational memory of *Valdez*, and determined to keep a log of decisions and issues that required attention as well as suspension. "I made a mental note. At some point somebody was going to come back and ask you some questions. And the logs would be the only thing we'd have to rely on because memory would be treacherous."

As at other district offices, the security of the federal buildings themselves became an immediate concern. "Everyone who was in a high building began wondering, 'why are we here?' We have planes landing here, we're in a tall building, it's a federal building, [the attack on the Alfred P. Murrah Federal Building in] Oklahoma [City] had already happened. The anxiety level was very high." As Admiral Brown and his staff stood up their Incident Command System (ICS), part of that planning involved preparations for moving the ICS command center to an alternate land location if the federal building in downtown Seattle had to be evacuated. If the Admiral required a floating command center, they would move to a tertiary site, a 378' Hamilton Class Cutter that would be sent from

PACAREA.

Contacts were initiated between the Thirteenth District and the Canadian Coast Guard at the O-6 and operational levels, both at the cooperative Vessel Traffic Service Center (VTS) that regulated the flow of traffic through the Straight of Juan de Fuca, and at the MSO. All commercial vessels in the district's waters were directed to anchor. Secure communications, required in order to talk with counterparts in the U.S. Navy, had to be quickly constituted.

After seeing to these requirements of command, control, and communications, Admiral Brown insisted that his people make sure that their families were safe and secure. "It was like the call of war," he remembered. "We need you here to do your job, but take a moment to check on your family and make sure they are okay. [You impart a feeling that] to protect my family I must do this. If I don't stand the guard, if we don't stand the guard, then who does? Warriors must war, so families can be free." Brown himself turned to familiar routines, recalling to his staff an observation from his country youth, when in troubled times his family would gather together. "Some of the women would cook, some would knit, the children would play, the men would gather around the truck to talk. If you watch when people get nervous some people will cook, some will knit, some people will rock, they go back to that thing that they feel comfortable with and know how to do. For me, it was a conscious effort to keep myself apart, not 'in' the events and activities, but back from those events and activities, watching them, and saying, 'what's next?' We need to establish communications. Fine. Give that to somebody. Don't sit there and get consumed by communications."

Admiral Brown led the assessment of port vulnerabilities and security needs in the Thirteenth, as well as the character of the enemy. "9/11 wasn't the first time for the twin towers; the *Cole* wasn't the first time. What you see is a very insidious, very resilient, very determined foe, one who is more than patient, one who constantly asymmetrically assesses you and waits. But time is equally dispersed to all of us. As it is a potential for my adversaries, it is also an opportunity for our allies." Brown challenged his field commanders to respond to him in writing about their assessments of the nature of the new war. Brown credited the input from these communications with structuring the district's response.

2. 9/11 in District Fourteen

In Honolulu, Hawaii, home to one of the largest concentration of armed forces facilities and flag officers in the country, it was the middle of the night. There, Rear Admiral Ralph Utley had just taken over as Fourteenth District Commander in June of 2001, after a tour at the Pentagon in the same billet filled on 9/11 by Admiral Jeffrey Hathaway.   Utley was awakened by a phone call from his Chief of Staff, Captain Thomas Yearout.

"Are you watching your TV?" asked Yearout, whose son had called him from Colorado to alert him to the attacks in New York.

"It's 3:52 in the morning," answered the Admiral dryly. "I'm not watching TV."

"You'd better turn it on, sir." Yearout also called the district Command Center, where personnel were already watching events unfold on television.

Admiral Utley asked Yearout to gather the senior staff, but thought better of it after he hung up the phone. "I immediately thought, 'Now that's dumb. You don't want to go to a federal building after somebody's just blown a big building up.'" Utley called the Command Center and redirected his staff to the Coast Guard facility at Sand Island, where computers were brought in to set up a crisis action center (CAC). Once at Sand Island, Utley closed the Prince Kuhio Federal Building in downtown Honolulu. Then, with his staff, he began to sort through the force protection and patrol needs of the district. By 0800 Honolulu time, the entire command structure of the district had been transferred to Sand Island. Utley's was the most direct and forceful response to an issue that confronted each district commander to a greater or lesser extent, and none more than Admiral Hull in the Ninth, who was forced to evacuate the federal building in Cleveland as the fourth hijacked airliner headed toward the city.

Phone conferences had already sprung up between Vice Admiral Riutta and his PACAREA district commanders, and Utley joined Admirals Thomas J. Barrett in Juneau and Errol M. Brown in Seattle in sharing whatever new intelligence came in as well as what each district was doing to lock down its port facilities and protect its forces. From his own staff, Ultey was looking for ideas, suggestions, what actions did they need to take and how were they going to accomplish it.

His biggest concern was the protection of the Port of Honolulu. Since 1939, the U.S. Navy had taken responsibility for the waters around Pearl Harbor, leaving the maritime and economic security of the Port of Honolulu to the Coast Guard. The 378' Hamilton Class Cutter *Jarvis* was in port in Honolulu, and scheduled to get underway on a counter-narcotics patrol. Utley asked PACAREA to chop the high endurance cutter to the district. Jarvis was directed to "stay nearby and be ready for whatever the district might need," and not venture more than one hundred miles from the port. *Jarvis* joined the rapid Navy sortie of its assets in Hawaii, and Admiral Utley asked the cutter to get a common operating picture of anything that was coming or going. "*Jarvis*," recalled the Admiral, "with its law enforcement mission and a rather large gun, [could] stop somebody from coming in who was trying to get in. I wasn't worried about that because I had *Jarvis* out there that could take care of that."

For Captain Yearout, it helped that the district already possessed a risk assessment tool that it used on a daily basis. "We were very much aware of terrorist problems, and had worked closely with the local Hawaii Emergency Preparedness Executive Committee (HEPEC)." HEPEC acted as a central committee for the anti-terrorism efforts of local, state, and federal agencies, including DoD. Especially in the aftermath of the terrorist attack on U.S.S. *Cole*, force protection had become a central issue within the massive military presence in Hawaii. Where other districts after 9/11 might have had to quickly build new relationships and partnerships with state, local, and federal agencies, Admiral Utley had those relationships in place already, along with port assessments of both Hawaii and Guam. And compared to Admiral Casto's Eighth District in New Orleans, which covered all or part of twenty-six states, Utley's Fourteenth District, albeit enormous in terms of ocean—over twelve million square nautical miles—covered just one state with seven counties and a well-coordinated civil defense apparatus and a few other Pacific islands. This same Hawaiian civil defense structure made a smooth pivot to homeland security after 9/11.

> We were on those conference calls on a daily basis with the district commanders talking about having to reach out to these people and those people. I was not exchanging business cards on the 11th of September. We knew each other. We already had the forum together to deal with this. We didn't have to stand up anything new.

Admiral Utley, however, recognized the immediate need for better and more secure communications, both between the Coast Guard and the Navy, and between Coast Guard units themselves. "This wasn't rocket science," said the admiral. "Everybody figured this out at about the same time. It was a blinding flash of the obvious." Such improved communications suites found themselves very quickly into the budget process. A similar effort got underway, at the urging of Vice Admiral Riutta, to combine the local Group and the MSO into a unified command, and to unify the district "M" and "O" staffs as well. "It was really easy to do here, because the CO of the group was a commander, and the CO of the MSO was a captain, so I didn't have to worry about who works for whom. This was not going to be a problem. This place was an excellent laboratory for doing this."

Admiral Utley worried that the Coast Guard, with its character changed fundamentally by 9/11, might let the opportunity for large-scale port security improvements slip by. After a robust Reserve call-up immediately after 9/11, the draw-down ordered several months after 9/11 cut deeply into the three-year strategy to bridge port vulnerabilities with secure capabilities. "We've gotten more respect. The challenge now … is to keep people interested in homeland security. We'll give you all the security you can afford, but we're going to have to grow… Historically, we think small. We see a large price tag on something and we say we can't ask for this much. [So] we trim things down to an unacceptable level internally before we even ask for it. That has to stop. We have to sit there and say 'we're going to do this right and this is the price tag.'"

3. 9/11 in District Seventeen

It was a bit before five o'clock in the morning in Juneau, Alaska, home of the Coast Guard's Seventeenth District, when the first plane hit the World Trade Center in New York. District Commander Rear Admiral Thomas J. Barrett, an Officer Candidate School graduate who had spent a year in Vietnam on board the 378' Hamilton Class high endurance cutter *Chase* and later attended George Washington University Law School, usually awoke around 0530. On 9/11, he got a call from his Operations Center asking if he had seen the TV or heard the news. Barrett flipped on his television and watched as the north tower burned. "This is Pearl Harbor all over again," he said to his wife. He got into his uniform and hurried from his house to his office.

Ten years of Admiral Barrett's Coast Guard career had been devoted to service in Alaska, a place that presented the Coast Guard with an intense operating environment, with long distances, terrible weather, and very little or no infrastructure. Routine search and rescue cases could become very treacherous very quickly. Barrett had focused his time as district commander on the requirement for the operational excellence that Alaskan waters demanded. The position of district commander involved both Coast Guard command as well as command of the U.S. Navy's sector of Alaska Command (ALCOM). Without the group commands common to other districts, the Ops Center in Juneau also tended to act as a tactical operations center for whole district.

As soon as Admiral Barrett arrived at his office, the district began recalling personnel and spooling up an Incident Command System. A call came in from Vice Admiral Riutta, who had just gotten off the phone with the Commandant. Riutta wanted the district to look over any potential targets and lock them down. First on the list was the Trans-Alaska Pipeline outlet at Valdez, followed by a substantial amount of international traffic flowing in and out of the state. Barrett very quickly attached the district's operational surface assets to the MSOs, which themselves possessed robust command and control communications. Air assets were attached to the MSOs on a mission by mission basis.

Prior to 9/11, there were no Coast Guard small boats at Valdez. The Coast Guard Auxiliary filled in immediately, racing its boat from the newly-created Auxiliary station at Whittier. Their actions did not go unnoticed by the district commander. "The Auxiliary loaded that safe boat onto a trailer and raced to Valdez where they launched within sixteen hours of the attack on the trade center. These unpaid volunteers went to extraordinary lengths to do it."

Admiral Barrett also touched base with the commander of ALCOM, an Air Force three star general, who was standing up a combat air patrol over the district. But Alaska was a long way from Manhattan, and being so far away from the incident's center of the gravity lent a certain lack of specificity to the threat analysis. As the morning progressed, that situation changed very quickly.

At about 1000 local time, a call came into Ops from the Air Operations Center at Elmendorf Air Force Base. Elmendorf was tracking a potential hijacked Korean Airlines (KAL) jet heading for Alaska. No one knew what was going on onboard the aircraft, but if it continued to behave erratically the Air Force might try to divert it to the airstrip at Coles Bay. This immediately put the Coast Guard Air Station at Kodiak on high alert. A large-scale search and rescue effort would be required if the airliner ditched or was forced down in a remote quadrant of Alaskan waters.

Barrett's Operations Center soon received a similar warning of a potential hijacking of the KAL flight from the FAA. No firm word of intent had been received from the airliner, but it was now turning toward Yakutat, so Barrett began to redeploy assets there. If the airliner was heading for Yakutat, that would push it toward Anchorage and Valdez, toward the very assets Barrett and his staff were most concerned with protecting. Soon after, the Ops Center received a report that the KAL flight had in fact been hijacked. As insane as it might have seemed, it appeared that someone was trying to crash a jetliner into the marine terminal at Valdez. What had looked like a localized attack in New York, one that had spread to Washington, was now apparently targeting strategic facilities in the American arctic.

In response to the potential threat from the inbound airliner, the Captain of the Port in Valdez made the decision to evacuate the port. All tankers would be pulled out. The Liquefied Natural Gas (LNG) Terminal and refinery at Nikiski were also closed, but Admiral Barrett did not feel that the same risk profile existed there. Nikiski was a remote port, away from population centers, a place where for the past twenty years LNG had been loaded onto the same two ships for delivery to Japan.

Valdez, however, was another story. Seventeen per cent of domestic U.S. oil production flowed through the port. Forty-eight per cent of the crude oil used by the State of California originated in Valdez. The potential economic disruptions of an assault on Valdez were obvious, as were the symbolic implications of any cut in the Trans-Alaska Pipeline. It was also in Admiral Barrett's mind what Iraq had done to the oil fields of Kuwait as the Iraqi army retreated in 1991.

By mid-afternoon of the 11th, the situation with the KAL flight became clearer. A pilot on the KAL flight had apparently mistakenly activated a hijack alert onboard the aircraft and it was this alert that led both the Air Force and the Coast Guard to scramble into position in case the alert became a direct threat to Alaska. The Air Force diverted the aircraft to an airfield in Canada, and the situation ended there.

With this nightmare out of the way, Admiral Barrett turned his attention to other potential Alaskan targets, and had only to look out his window. The season was far advanced, but cruise ships still called at Juneau until the end of September. A half dozen of these behemoths, each with two or three thousand people on board, were moored at that moment along the Juneau waterfront or cruising Alaskan waters. Each now became a potential *Achille Lauro* or, worse, a floating World Trade Center. Just as Admiral Hull immediately began to monitor bulk carriers from his Ninth Coast Guard District headquarters in Cleveland, so Admiral Barrett in Juneau began to track all of the cruise ships in District Seventeen waters as a prelude to ratcheting up security around them.

> We never thought of [the 9/11 port security response] as a crisis. It wasn't a new mission... It was an issue in Vietnam with transport vessels and explosives vessels. That mission was always there. I think we had a lot of people not used to doing it, but our people stand up very well. But we had disarmed the "M" folks. When I came into the Coast Guard, all the MSOs had people who were weapons qualified—they had weapons lockers that were part of their normal capability. Then we took all that away. Right after 9/11 we went right back there. Folks have got to be able to put out armed boarding teams. We don't always have operational units with LE teams available. We went right back to re-qualifying folks at the MSOs with weapons. One of the early questions I got from MSO Anchorage was 'We're happy to do this. We know we have to do this. But is this going to be a short fuse thing or am I going to invest all this time and energy and in six months someone is going to say it's not the right answer?' I said 'As long as I'm up here we're going to do it.' We had to get very quickly back to being a more armed, better armed, better trained Coast Guard.

Other missions emerged after 9/11 in Alaska, such as escorts for U.S. Navy submarines as they calibrated their noise levels off Ketchikan. In Valdez, law enforcement detachments were placed on board 110' Island Class Cutters and told to have their weapons loaded and ready. "It's an awareness mission, a presence mission, a deterrence mission," recalled Admiral Barrett, "but ultimately it's a stop 'em mission." In the event of a terrorist hijacking incident, the Coast Guard response would be conditioned by the nature of the incident. A federal hostage-taking scenario would be coordinated by the FBI, with the Coast Guard in a supporting role. A similar incident on a state ferry would be handled by the State

Police, again with the Coast Guard in a supporting role.

Other scenarios, however, fell between the cracks and had to be rethought. 9/11 forced Admiral Barrett's crews in Alaska to look at the previously innocuous as the potentially sinister. About a week after 9/11, a Coast Guard air patrol over Valdez noticed a boat anchored behind an island. It was an aluminum boat, about twenty-four feet in length, with twin outboards.

> It was just anchored there [and prior to 9/11] we wouldn't have blinked at that at all. Don't know whose skiff it is, but it's inside the entrance, and we've got oil tankers transiting in and out of here, where they have to slow down. So what do you do? We couldn't locate who owned the skiff. So we sent a buoy tender in there and collected it. We would never have done that before 9/11. It turned out that it was the fishing skiff of a Fairbanks fellow, who would fly his floatplane out there, fish, then fly home. And he's been doing that for years. But the filter through which we looked at stuff like that became very different.

Chapter Ten

9/11 and the Reserve and Auxiliary: PSU 305 in New York Harbor and Guantanamo Bay, Cuba

Damage Controlman Third Class Eric Bowers and Boatswain's Mate Third Class Michael Walker of the Coast Guard Reserve's Port Security Unit 305 had known each other since first grade. In civilian life, they were both firefighters for the District of Columbia. They went through boot camp together, were hired by the fire department and went to firefighter's academy at the same time. Just after the attack on the Pentagon, all DC firefighters were recalled. Bowers found himself as part of the first DC fire company on the roof of the Pentagon, cutting into that roof to try and vent the fire out of the biggest office building in the world. "It was a tough roof to breach," said Bowers. "Layers of slate, wood, and then concrete. We were up there a long time. I could pick up and look at American Airlines flight magazines fluttering around, from the 757 that hit the building. There were body bags everywhere, which you see all the time as a firefighter, but never on that scale."

Petty Officer Walker found himself in the blazing interior of the Pentagon. "A lot more of the building was burned on the inside than appeared from outside. It looked as if the fire blasted down the hallways. Because of the lay-out of the building, even though I'm on a truck company, for all those hours we were doing engine company work, pulling hoses, dragging hoses. There were offices with thin wooden doors, but because the doors were closed the offices were fine. Then there were offices with open doors where the people never got out of their chairs, where people were melted into the linoleum."

Bowers and Walker had little time to rest that night. And early the next morning, on 9/12, they received phone calls from their Coast Guard Reserve port security unit. They had been mobilized and ordered to New York Harbor.

*   *   *

Under Title 14 of the U.S. Code, the Secretary of Transportation had the authority to mobilize a limited number of Coast Guard Reserves for up to sixty days in any two-year period. Admiral Loy's request to Transportation Secretary Norman Y. Mineta for authority "to involuntarily recall the selected reserve of the United States Coast Guard" is interesting for several reasons. Besides the implication that Loy was asking for the entire Reserve, the memo to the Secretary also reflects as well the initial confusion of that morning. "Terrorist elements appear to be involved in attacks on the World Trade Center in New York City, the Pentagon in Virginia and the State Department in Washington, DC [italics inserted]."

Four days after the attacks, the president declared a national emergency, effectively transferring the recalled Reserve from Secretary Mineta's Title 14 authority, to the president's Title 10 authority. Title 10 gave the president sweeping powers to mobilize reserves for different levels of national emergencies, ranging from a low of 200,000 reserves for 270 days, to a full mobilization of all reserves for the duration of the conflict plus six months. After 9/11, the Coast Guard Reserve was mobilized under Title 10 for an interim step, called 'partial mobilization,' in which the president exercised his authority to call up to one million reserves for up to two years.

Prior to *Desert Shield/Desert Storm*, the Coast Guard Reserve had been mobilized only twice between 1973 and 1990, for Midwest floods in 1973 and during the Cuban Mariel Boat Lift in the summer of 1980. In August of 1990, 1,649 reserves were called to active duty to support *Desert Shield/Desert Storm*. In the decade since the Gulf War, streamlining had shrunk the Coast Guard Reserve from 12,500 to an effective force of barely 7,500. By 9/11, most of those remaining 7,500 reserves had long since transitioned from the Cold War reserve unit structure, which had existed to provide training for mobilization in the event of a general war in Europe, to individual day-to-day augmentation of the Coast Guard's active duty and civilian workforce. In fact, the Coast Guard was proud of its 1996 decision to integrate the Coast Guard Reserve into a "Team Coast Guard" concept. But it proved the wrong construct post-9/11.

Even so, elements of the Reserve had been called to active duty three dozen times in the 1990s, mostly to assist in the Coast Guard's responses to hurricanes (fifteen mobilizations), floods (six mobilizations), and tropical storms (three mobilizations). With the exception of 271 reserves called to active duty to support Operation *Uphold Democracy* in the fall of 1994, military operations of the Reserve were negligible.

The day before 9/11, 983 Coast Guard Reserves—thirteen per cent of the total reserve force—were already on active duty in various jobs. In effect, any reserve who wanted to be on active duty prior to 9/11 already was. The reserves mobilized after 9/11, were true reserves, people who held down regular civilian jobs. Many of these were police, firefighters, and EMTs, the very jobs now recast as vital to homeland security.

Mobilized reserves were of primary concern not only to the district commanders, who wanted more resources to protect their growing critical infrastructure lists. They also immediately raised a flag at the Chief of Staff's office, which had to find the money to pay them. A memo circulated through Coast Guard Headquarters less than a week after 9/11, expressing the discomfort of the Chief of Staff's office with "the reserve number and associated costs." The memo strongly hinted that the mobilization be capped at no more than 3,500 reserves.

At the height of the post-9/11 mobilization, 2,751 reserves were called to active duty. Added to the 953 already on active duty on 9/10, this meant that nearly fifty per cent of the Reserve was on active duty between mid-September and mid-November of 2001. Put another way, during these crucial eight weeks, it was doubtful that the U.S. Coast Guard had any Reserve left with which to respond if another attack crippled another American port.

By November 1st, Admiral Loy was searching for ways to draw down "the largest surge of Coast Guard Reserve forces since World War II." The Service simply could not financially sustain a long-term reserve mobilization. Loy asked for "adjustments" and "fine-tuning," which in practice produced a swift demobilization. With the exception of a small number of reserves kept on board for specific tasks, the majority of reserves had returned to civilian life by September, 2002, the end of the 2001 fiscal year.

The hurried call-up and equally rapid draw-down left some reserves in a kind of quasi-mobilized limbo. For the joint Coast Guard-Navy Harbor Defense Command Unit (HDCU) 201, a command and control element of the Navy's coastal warfare community lead by Coast Guard Reserve Captain Joanne F. Spangenberg, the fall and winter of 2001-2002 brought an exasperating series of deployment alerts and cancellations. Originally called up on November 18, 2001, to perform homeland defense, the unit was still mobilized, and still waiting for deployment orders, in early April, 2002. A potential mission to the Mediterranean in January had been scrubbed, leaving personnel packing and repacking pallets of gear, training with new equipment, and marking time.

Part of the ambivalence seemed to stem from a hurried realization that the few harbor defense and port security units available to the Coast Guard were not in any way numerous enough to guard the thousands of sites now listed as critical national infrastructure. Yet the feeling of time and opportunity slipping away was very real. Captain Spangenberg noted that the American flags carried on so many civilian automobiles after 9/11 had faded or been discarded altogether by spring.

*   *   *

For the Coast Guard Auxiliary, 9/11 proved something of a coming out party. An all-volunteer nautical service organization of approximately 34,000 men and women with an average age of 57, the Auxiliary before 1996 was engaged primarily in teaching boating safety courses, conducting Courtesy Marine Examinations for recreational boaters, and providing weekend search and rescue services for recreational boaters. The Coast Guard Auxiliary Authorization Act of 1996 had modified the group's charter, allowing the Commandant to use auxiliarists in any Coast Guard mission except direct law enforcement and military combat. Auxiliarists after 1996 could be found filling in for both active and reserve forces during surge operations during natural disasters and environmental responses. The Auxiliary had taken on some non-emergency search and rescue work as well.

9/11 dramatically elevated the profile of the Auxiliary, as civilian volunteer auxiliary boats and aircraft conducted both surface and air patrols and assumed search and rescue standby postures at many small boat stations. This backfilling of the active force allowed the active duty small boat force the latitude to redirect Coast Guard small boats toward port security and homeland defense. On 9/11, auxiliarists distributed face masks and gloves, along with food and water, to rescue personnel at Ground Zero. On the water, Coast Guard Auxiliary boats from Westchester County ran security patrols around the George Washington Bridge, and those from Sandy Hook did the same at the Verrazano Narrows Bridge. Auxiliary boat crews backfilled for Coast Guard search and rescue crews at Stations New York, Sandy Hook, and New London, taking over as communications watchstanders in New London.

At Tarrytown, New York, legally blind Auxiliary radio operator Mike Coffey worked continuously to monitor patrols. Coffey later received the Award of Operational Merit for his work. Another auxiliarist and licensed psychologist, Dr. Janice Jackson, served as a backup to the Critical Incident Stress Management Team throughout the day on September 13th. In the six months after the attacks, the Auxiliary contributed nearly a quarter of a million volunteer hours to the Coast Guard's 9/11 surge. Auxiliary surface and air patrols alone amounted to 10,139 people conducting 7,454 sorties over 53,910 hours. The total number of Auxiliary volunteers after 9/11 exceeded 27,500, or over eighty per cent of the total Auxiliary force.

Vice Admiral Ray Riutta, Commander of the Coast Guard's

Pacific Area during 9/11, said the Service would not have been able to accomplish what it did after the attacks without the backstop of the Reserve and Auxiliary. "Much of the expertise we needed resided in the reserve force that we brought on board, and much of the relief we needed, everything from driving to manning radios to actually doing patrols and delivering Sea Marshals, we got from the Auxiliary, which stepped up to the plate huge. Had we lost the Reserve, as was threatened a few years ago, or not had the Auxiliary, we would have broken our force within a matter of months. Even as it was, we stretched everyone to their limits."

*   *   *

One group of Coast Guard Reserves kept either on continuous stand-by when not on active duty after 9/11, were those reserves formed into Port Security Units (PSU). Comprising a bit more than ten per cent of the total reserve force, PSUs were also the last pocket within the Coast Guard Reserve that retained much of the pre-streamlining reserve unit structure. PSU 305, for example, was a port security unit consisting of 135 Coast Guard Reserves and five active duty personnel, based at Fort Eustis, Virginia, with a full complement of its own reserve officers. The PSUs were also unique in that they wore battle dress uniforms with identifying insignia for each individual unit, and their highly-visible, well-armed Boston Whaler gunboats were painted with gray slashes rather than the usual orange slash.

The overtly-military units trained to deploy outside the continental U.S., to offer port security in foreign harbors used by U.S. forces in times of both war and peace. As such, units like PSU 305 were not unfamiliar with the Middle East, having supplied port security for Operation *Bright Star*, a two-week exercise in Egypt in 1999.

The commanding officer of PSU 305, Commander Robert W. Grabb, a former enlisted Marine Science Technician who began his career on some of the Coast Guard's last Ocean Station weather patrols of the 1970s, had been a plankowner of 305 as its operations officer since its formation in 1994. Like the personnel of the National

Strike Force, Commander Grabb's people tended to keep one eye on their jobs and another on CNN. In the case of the PSU, a harbor security situation anywhere in the world carried the potential for a unit deployment. "On 9/11," remembered Grabb, "even before we got the recall, people assigned to the unit were already calling in... What we advertise is that, within 96 hours of receiving a recall notice, we can have everyone recalled, fully loaded, fully palletized, and have everything on the tarmac ready for airlift, wheels-up, anywhere in the world... So our people pay attention to world events."

After their withering day at the Pentagon, during which they never saw each other, DC fire fighters Eric Bower and Michael Walker were awakened on the morning of September 12th by phone calls from PSU 305 recalling them to active duty. Much of the unit's boat division was already mustered at their home base at Fort Eustis, Virginia, having come in on Monday for a few days of training. The unit's Boston Whalers, fitted out with twin 175-horsepower outboards, were loaded on flat bed trucks and sent north to New York.

At noon on Thursday, the personnel of PSU 305 moved out from Fort Eustis, and by that evening encamped temporarily at the gymnasium of Activities New York and came under the command of Admiral Bennis. Once in New York, PSU 305 eschewed the facilities of the Station New York, in favor of staging their Whalers at the Military Ocean Terminal in Bayonne, New Jersey (MOTBY). From MOTBY, 305's boats escorted into New York Harbor and eventually stood guard over the U.S. Navy hospital ship *Comfort*, moored along the Hudson River north of the area of the World Trade Center.

Deploying to an American port to guard Americans was a unique experience for 305. The close proximity of civilians and civilian assets created new challenges, especially with regard to use of force. The .50 caliber M2 machine guns on board the Whalers had a range of over four miles. In a crowded harbor on choppy waters, such force had to be tightly controlled. Lieutenant Commander Lee A. Handford, 305's executive officer, noted that the unit "trains on clearing fields of fire, and ensuring to the maximum extent possible that we don't have 'friendlies' behind a target. That's not always possible. We are designed as a warfighting force. [Our] priority is protecting a high-value U.S. asset." The unit's self-contained armory was eventually placed by a crane inside a fenced handball court, where gunner's mates lived behind the fence and gate.

The half-hour transit from MOTBY to Manhattan began to wear down the boat crews, who were more used to enforcing set security zones than engaging in harbor or river patrols. Through mid- to late-October, dropping temperatures made life on the open fiberglass gunboats less than a pleasure for a unit set up for continuous operations. "The PSU watch, quarter, and station bill, is set up to do three sections indefinitely," said Lieutenant Commander Handford. "Eight hours on and sixteen hours off. That is a grueling pace in itself, but when you add to that briefs and debriefs, and transport to and from launch points, it turns into a ten-hour watch. In New York, it rapidly turned into a twelve-hour watch." Or, as the Operations Officer, Lieutenant Commander Karl Leonard, put it, "everyday is a Monday for us."

After 45 days on patrol in New York, the members of PSU 305 were released from active duty, to enjoy about two months at home before they were called to active duty again. Originally scheduled for a six-month deployment to the Middle East beginning in March, 2002, they found themselves instead deployed in early January to Guantanamo Bay, Cuba. PSU 305 was attached at Guantanamo Bay to Joint Task Force 160 (JTF 160), a force of primarily Marines of the 2nd Force Service Support Group from Camp Lejeune, North Carolina, under the command of Brigadier General Michael R. Lehnert. Lehnert's Marines took care of land security for the antiterrorism detainee operations, but the Marines needed waterside security as well. Lehnert was familiar with the Coast Guard Reserve PSUs, since they trained at his home base at Camp Lejeune.

The Defense Department therefore, asked the Coast Guard to provide waterside security for Camp X-Ray, where Al Quaeda and Taliban captured in Afghanistan were being confined. Together with the U.S. Naval Reserve Mobile Inshore Undersea Warfare Unit 208 from Miami, the Coast Guard and Navy units formed JTF-160's Joint Maritime Patrol Group. PSU 305 developed good working relationships with both the Marines of JTF 60 and the Commander of the Naval Base. A rather abrupt transfer of regime from the Marines to a U.S. Army National Guard command led to the reinvention of several wheels and a decided lack of familiarization with the new Army faces.

After five months in Guantanamo Bay, the members of the unit began to realize that the new global war on terrorism offered every indication of involving them in larger and more frequent mobilizations. Part of these feelings stemmed no doubt from the usual grumblings of any deployed military unit, especially an operations intensive sub-unit like PSU 305's boats division that ran twenty-four hour patrols, while the occasional Coast Guard cutter ambled into Guantanamo for morale time. At such times it was easy to imagine two distinct Coast Guards, one 'blue,' one 'green.' But another part seemed to come from the fact that 9/11 literally hit most members from the clear blue sky. These were reserves who had perhaps grown accustomed to believing that *Desert Shield/Desert Storm* had been the last great American military mobilization. Lieutenant Commander Leonard became acutely aware of such feelings on the part of his personnel, and sought to counter them at every turn.

> For our personnel, I think there needs to be a real realization of what you're signing up for. I can't speak for recruiters, I know that even before 9/11 there was pressure to increase the reserve force. I don't know what recruiters are telling people or how much pressure they're under to sign, sign, sign. But I'm finding out that a lot of our members were just not prepared financially, didn't realize the possibilities. I think for everybody stepping foot into this unit, they need to know that it's not a matter of "you may," you will be deployed. Sometime in your career in this unit you will be deployed for a long time, and it will be at your current rank salary.

Even for Leonard, however, it was difficult to imagine why someone would take lower level enlisted pay for six months in a relative hellhole like Guantanamo Bay, when they could be assigned instead to an MSO instead and have the possibility of going home every night. The challenge of recruitment and retention of reserve port security personnel, he believed, would continue to be problematic without some accompanying carrot to go with the certain stick of long-term overseas deployments. The leadership of the unit was braced for a flurry of requests for transfer when they returned to Fort Eustis.

An as operations officer, Leonard saw an even bigger challenge in getting the Service to understand the nature and capabilities of its port security units and their range-limited, wide open, fiberglass gunboats. PSU 305 had adapted almost instantly to in many ways a fundamentally different mission in both New York Harbor and Guantanamo Bay. But the unrelenting character and tempo of those missions had taken a toll on personnel, especially on those petty officers with less than ten years of experience, the very cadres the Service needed to maintain unit continuity in the decade after 9/11.

For firefighters and Coast Guard Reserves Bowers and Walker, they had seen and served in three of the major sites in the global war on terrorism. They had responded to the terror attacks in both New York and Washington, and had served overseas to contain captured terrorists. In the 276 days separating 9/11 from June 15, 2002, from the moment they left their homes to put out the fire at the Pentagon to the day they departed Guantanamo Bay to return home, Bowers and Walker and their port security unit had seen 226 days of active duty. Sitting in the rain in Guantanamo Bay, Walker saw little to be enthusiastic about, except the thought that someday they would be able to look back with intense awareness of what they had accomplished. "When we're older and we have kids, they'll bring their books home with writings about these events, and we'll be looking through the pictures and, I don't know, maybe it will have more effect on me then." He and many others looked forward to simple things they had missed: green grass, flowers, a favorite pickup truck.

PSU 305 returned to its home base in Virginia in mid-June, 2002. Discussions were underway regarding possible U.S. military action in the Middle East. Within a few days, the unit had begun to repack their gear in preparation for the next time they would be called. No one was now in any doubt that the next call would not be long in coming.

Conclusions:

The Coast Guard and the New War

"Know the nature of the war. This is not a classic example of force on force. Striking soft, innocent, civilian targets with no conscience and no restraint is so profoundly different that it almost defies strategy. But if that's what you believe the nature of the war is, then it guides how you act and how you prepare."
> -Rear Admiral Errol Brown,
> Thirteenth District Commander during 9/11

The effects of 9/11 on the U.S. Coast Guard were dramatic, both immediately and in the long-term. On the morning of 9/11, the Coast Guard was struggling through seemingly annual budget travails as one of many diverse constituents of the Department of Transportation and the country's "fifth armed force." By the end of the day on 9/11, as the Coast Guard demonstrated its unique value to the American people in very public ways, the Service began to emerge from both its bureaucratic and military obscurity. On the strength of its wide-ranging maritime legal authority, the Coast Guard had moved to secure every strategic port in the United States, and initiate a long and intensive process to secure the nation's maritime infrastructure. By right of its standing as an armed force, the Service had mobilized nearly 3,000 reserves almost immediately, and coordinated the maritime evacuation of Manhattan. By the end of the day on 9/11, both immediate and long-term funding was in the works such as the Service had not enjoyed in its entire history.

A year later, the Coast Guard was at the forefront of nearly every discussion of national security. The Service entered the everyday consciousness of the American people with greater impact than at any point since the Second World War. A year later, 9/11 was beginning to be seen as a hinge in the history of the Service rivaled only by the creation of the modern Coast Guard in 1915 or the formation of Alexander Hamilton's original Revenue Marine in 1790. The "fifth armed force" of the morning of 9/11 had by the end of 2002 become the core of maritime homeland security. The performance of the Coast Guard's 9/11 Commandant, James Loy, was such that upon retirement from the Service he became Chief Operating Officer of the new Transportation Security Agency and, not long thereafter, the Director, a key figure in the new Department of Homeland Security.

A year after 9/11, the effect of the terrorist attacks had also reached deep into the culture of the Coast Guard itself. Nearly every program within the Service was looked at anew, within the context of the security of the homeland. New, primarily active duty, Maritime Safety and Security Teams (MSST) were created specifically to counter a range of threats to domestic ports, even as many members of the traditionally overseas-focused Reserve Port Security Units wondered if this deployment structure wasn't constructed backwards. Some wondered why no one thought to recreate the old Coast Guard Reserve port security structure, which drew its members from local police and fire departments and had functioned to protect domestic ports throughout the Cold War. With the fall of the Berlin Wall in 1989 and the Yeltsin Revolution in Russia in 1991, with the 'streamlining' of the U.S. federal government in the 1990s, most of that structure had vanished. No one missed it until the morning of 9/11.

Over the years, Coast Guard Reserve port security units had gradually self-selected for certain types of civilians: state and federal employees, and law enforcement and fire fighters. As the executive officer of PSU 305, Lieutenant Commander Lee Handford, remarked, besides their obvious useful skills, "they're the ones who can take the extra drills that we get, not to mention the extra deployments." Increased training requirements and weapons qualifications, not to mention such seemingly unrelated tasks as public affairs events like parades ("it's easier for the Coast Guard to send a small boat and a couple of guys than a white hull") had dramatically increased the time commitment for these specialized reserves, even before September 11th.

Yet it was these civilians—law enforcement and fire personnel—most dramatically impacted by 9/11 and the new demand it created for local homeland security regimes. Extended and repeated out-of-country (OCONUS) deployments for personnel already vital to homeland security created a whole new set of issues that strained employer support of the Reserve. If the six-month crisis deployments, like those to New York and Guantanamo Bay, continued, along with the increasingly routine six-month Southwest Asia deployments, the personnel of the six Reserve PSUs would eventually reach breaking points in civilian careers. 9/11 seemed to turn the philosophy of mobilizing the Reserve for "surge requirements" into an anachronism, as the country moved toward a kind of permanent counter-terrorist security posture. As Lieutenant Commander Karl Leonard, operations officer for Port Security Unit 305 during 9/11 and a police division commander in civilian life put it:

> We all got called up during Desert Storm, and I know stories of police officers that were fired, terminated, when they left. I missed out on a promotion. We had a lot of that negative stuff attributed to it. When we came back we had to hire attorneys and fight for our jobs. This time, I think because they struck at the heart of America, the support was overwhelming. Go, we're paying you benefits, we're paying this, we're paying that. So the police on that front have really been taken care of. The other side of it is, September 11th made domestic terrorism a priority for local law enforcement. They now have to step up programs in their towns, cities, villages, and counties. They need to implement stronger watches, more personnel here and there, so the task and burden on them has really increased tremendously. And while we're increasing their responsibility we're taking away their officers. My department alone had forty-five officers called up, out of 500.

Post-9/11 proposals to return the Coast Guard Reserve to Cold War levels, therefore, came at a time when recruiting and retention were becoming more and more problematic.

9/11 also occurred in the middle of the largest acquisition of operating platforms in the history of the Service, the Integrated Deepwater System (IDS). On 9/11, IDS was led by Rear Admiral Patrick Stillman, charged by Admiral Loy to look years and decades into the future and piece together a combination of technology and people for a Service that defied simple categorization. A former Captain of the Coast Guard's 295' training tall ship, *Eagle* (W327), Admiral Stillman was someone who operated comfortably with a palm pilot in one hand and a biography of Nelson in the other. He seemed by training and temperament as an officer who could extract the essential elements from two centuries of Service history in order to design a Coast Guard fleet of the future.

In a world dominated by complex technologies and over-whelming streams of data and intelligence, Stillman looked to the *Eagle* as both ship for the Service and metaphorical ballast for its sailors. "All of us are afforded the opportunity to sail on the proper tack in life. *Eagle* is a terribly complex entity on the surface, but in reality it's a very simple enterprise. You have twenty-three sails, but ninety per cent of your drive is tied to six of those sails. It provides adept simplification, and mandates the need for self-assessment, courage and humility. When you get up on the royal, when you're hauling on the braces trying to bring the yards around, you can't do it absent somebody standing next to you. It's a cradle of communi-ty in many respects." Deepwater, the functional equivalent of the NASA's *Apollo* program, could be seen as the cradle of the future Coast Guard community, offering not only a new fleet of ships and aircraft for a new century, but a recreated sailor as well.

The name of the project itself had caused some anxiety, even within the Service, as it offered the impression that the technologies would be deployed more than fifty miles offshore, on patrols of the U.S. Exclusive Economic Zone (EEZ). In fact, many of the platforms being replaced—helicopters and smaller cutters like the 110' patrol boat—were coastal assets as well. Proposals solicited from defense contractors in June of 2001 were due at the end of September. Just as these proposals were scheduled to arrive at Coast Guard Headquarters, the attacks of September 11th occurred. In fact, Admiral Stillman had just left a Deepwater briefing with Admiral Loy as the first plane hit the World Trade Center. Against the backdrop of the 9/11 port security crisis, formal evaluation of the proposals began on the first of October. "With 9/11, and the impact of that event on the long-term missions and requirements of the Coast Guard, it was important to step back and take a look at the constructs and foundations of the acquisition to ensure that the requirements that we had specified were still germane and current and not in need of significant modification."

Using a metaphorical approach to Coast Guard operations, Stillman reasoned that the Service had engaged in the surveillance of the coastal and offshore environments ever since Alexander Hamilton's first Revenue Marine cutters slid down the ways. This surveillance was always followed by the detection and classification of contacts, and then by decisions over how to prosecute those contacts. IDS was Stillman's bridge between Hamilton's 18th Century Revenue Marine and Loy's 21st Century Maritime Domain Awareness. "Because we had this common historical strain of task sequence, tied to multiple missions, [after 9/11], industry was afforded the opportunity to increase our operational effectiveness as it pertained to our ability to surveil and prosecute contacts." Seen in this historic context, Stillman saw little need to change the established Deepwater model after 9/11, because maritime domain awareness technologies had been part of the project from the start.

> The interagency task force on the roles and missions of the Coast Guard that was conducted in '99, and hit the streets in early 2000, spoke to the Deepwater acquisition, and encompassed the nature of asymmetric threats. Terrorism was a concern that we needed to be attentive to over the next decade and in terms of the 21st century Coast Guard. Because that notion had already been embedded within Deepwater from the start, the project was on solid ground after 9/11 and [that event] didn't mandate any significant change.

Deepwater presented the Service with a somewhat radical notion of its own future, and Stillman expended much thought in defining this notion. The Coast Guard was a small Service, and as such tended to mute the differences of rank and privilege inherent in most hierarchical organizations. The Coast Guard tended to be flatter than other Services, both in terms of its culture and its structure. It was one reason why you could find a young petty officer in charge of the first Coast Guard asset to make its way across New York Harbor after the World Trade Center attacks.

9/11 did, however, amplify some of the new Deepwater elements that had been on the margins of the radar screen prior to the attacks. Captain Dan Deputy, force manager for the cutter fleet and chair of the National Fleet Working Group, was in the middle of discussions as to what weaponry the new cutters would carry when 9/11 occurred. The multi-mission nature of the new cutters, combined with the multi-mission character of the Service itself, challenged traditional ideas about platforms and crew. Deepwater envisioned a new Coast Guard fleet where several different configurations of weaponry as well as crew would be available on a single cutter—for example, a medium caliber gun with Coast Guard crew for a law enforcement mission, a missile launcher and Navy crew for a naval warfare mission, an intelligence team for a listening mission—and the mix of those elements would be contingent on the national security task the cutter was sent to perform and its operational requirements.

Deputy's staff looked at a host of elements the cutter force would require in a post-9/11 world. These centered on new technologies for biological, chemical, and radiological detection, but also included a new evaluation for the concept of 'optimal crew size.' Optimal manning worked well on large cutters with predictable underway schedules, less well for buoy tenders and other smaller cutters with often irregular hours away from the docks. As the ultimate irregular event, Deputy believed that 9/11 significantly changed the established concepts of regular hours and program schedules.

> 9/11 has brought up the fact that all of this optimum manning is a great concept on paper. [We had developed] cutter support teams (CST) and maintenance facilities on our newest buoy tenders—the 175s and the

225s. They set up with a smaller crew with these maintenance facility teams that are on the shore, yet they were supposed to come on board and do all this work yet not be a part of the crew. They would be managed by a shore facility. Over time, it didn't work out as well as maybe it should have… A 175 goes out when the weather's appropriate for buoy work, or they work a buoy that's discrepant and work buoys that are scheduled. But they can't guarantee that they're going to go out on Tuesday and come back on Thursday… The ships were too optimally crewed, and we realized that right off the bat.

The compromise eventually adopted was to give the support staff to the commanding officer of the cutter, who then decided who remained ashore and who went to sea. "The ships that were under the [new] CST concept were able to meet the homeland security/port security needs a heck of a lot better than the other cutters," recalled Captain Deputy. "Like the [Keeper Class coastal buoy tender] *Frank Drew* [(WLM-557)] that came up the Potomac. Since they had a CST, they were able to rotate people out, go home, come back… There are a lot of advantages to having the CO able to make those decisions."

The multi-mission character of the Service led to unique personnel challenges, as well, and Admiral Stillman among others devoted many hours of thought to the Service's culture and structure, and how those components formed organizational competence. "It's not at all uncommon for a captain and a seaman to be interacting on a daily basis," said Stillman, "so that over time a comfort level tends to develop. Maritime safety and maritime security have always been fundamentally part of the mores of the organization, and these have attracted individuals accordingly."

Such unique personnel characteristics meant that any program as sweeping in its impact as Deepwater would invariably cut deeply across Coast Guard culture. Deepwater did not seek to add a cutter here or a patrol aircraft there, but rather to introduce a whole new system of operation that would cross platforms, people, and performance. By several definitions, this meant that all the new technologies would require a new kind of Coast Guard sailor as well. If the Service truly wanted to improve its operational effectiveness, while controlling and reducing the associated costs of operations, then prudent managers had to look at the cost of people, which comprised two-thirds of the operating budget. For Admiral Stillman, this meant that the Service in the years to come had to be more intelligent, more highly trained and, contrary to many post-9/11 prognostications, smaller. Therefore, effectiveness in the operational environment meant leveraging more sophisticated technology to make fewer people far more efficient. Part of the solution, Stillman envisioned, would be patrols of the EEZ with unmanned vehicles, both aerial and surface, with logistical support contracted from outside the organization.

> The reality is that thriving organizations in both the public and private sector are those that are adept at re-engineering, because change is fundamental to the workplace, and if you can't adapt expediently to that change you will suffocate and suffer the consequences. For us, the mandate for the effective delivery of public services forces us to re-engineering constantly, constructively, and continually. I don't embrace the dictum that to do it right you have to do it yourself. I think that is, frankly, passé. We've had the sea as a teacher for more than 200 years. I think it's promoted a sense of humility and artistry that permits people to be can-do and optimistic in their deportment. If we operate under the dictums of duty, honor, and respect, we have every reason to be optimistic.

Even as Deepwater proceeded, 9/11 clouded the operational near-horizon for the Coast Guard. Exactly what missions would all these new and more sophisticated platforms perform in 2020, or 2010, or next year for that matter? The Service was legendary for an amoeba-like ability to gravitate toward the mission *du jour*, which was generally, if not cynically, equated with the mission promising the best short-term budget stimulus. Vice Admiral Thomas J. Barrett, Commander of the Seventeenth Coast Guard District during 9/11 and promoted to Vice Commandant in the aftermath, could not envision the Service shrinking in numbers, especially if the number of missions the Coast Guard was asked to perform continued to increase each year. "One of the problems we had with the last drawdown is that they didn't change the mission suite and they didn't change the equipment," he noted flatly. The implication was clear, and had often been expressed in one of Admiral Loy's favorite euphemisms: the logical outcome of doing more with less was doing everything with nothing.

Admiral Barrett remarked on the raft of new MSSTs, on new inspection requirements, on more maritime domain awareness requirements, on the continuing refinement of search and rescue technologies and techniques, and on 5,000 new active duty forces and an equal number of new reserves. All these would require more, not fewer personnel.

Fourteen months after 9/11, on Monday, November 25, 2002, President Bush signed into law the legislation creating a cabinet-level Department of Homeland Security (DHS). The Coast Guard would become the largest component of the new department, a separate and distinct, yet integrated operational element, akin perhaps to the relationship of the Marine Corps to the U.S. Navy. The reorientation of the U.S. Coast Guard from the Department of Transportation to the Department of Homeland Security, occurring simultaneously with the development of Deepwater technologies and the integration of both the Service's command and control on the one hand and its marine safety and operational cadres on the other, foreshadowed many years of demanding and perhaps wrenching changes throughout the organization. The internal challenges inherent in joining DHS were multiplied by the merger of the Coast Guard with the twenty-one other federal agencies that would form the DHS. The decade ahead of intense policy challenges facing Coast Guard leadership was eclipsed only by the ever-increasing pace of the Service's immediate operational tempo.

The Coast Guard's constant response mode of operation highlighted the Service's lack of mission continuity. It was difficult to produce thoughtful officers when the mission set changed so rapidly, offering little time for strategic analysis, contingency planning, and dissection of lessons learned from previous operations. Whether it was *Exxon Valdez* or the Mariel Boatlift, the response mode of operation left little room for proactive doctrine-building. It was an insecurity of core mission known to no other American armed force.

Captain Joel R. Whitehead, the Chief of Staff for the Eighth District during 9/11, remembered his first tour at headquarters in the mid-1980s, when his work in marine safety had included a heavy dose of anti-terrorism. He recalled how sharply he and now-Admiral Stephen W. Rochon were focused on maritime terrorism as a direct consequence of the hijacking of the cruise ship *Achille Lauro* and the subsequent murder by Palestinian terrorists of disabled 69-year-old American passenger Leon Klinghoffer. "We stood up a branch just to track maritime terrorists. That all went by the wayside in the last fifteen years. That whole focus disappeared. By '93 or '94 we had dismantled all that. It wasn't a priority anymore."

Others thought that perhaps 9/11 would lead to stable long-term missions, even as those missions adapted to the shifting tactics of terror. It was the reason why Admiral Paul J. Pluta saw 9/11 as bigger than either the Mariel Boatlift or *Exxon Valdez* in terms of Service impact. Unlike previous crises, 9/11's effects were felt not just within the U.S. but around the world, and the scope of the legal architecture framing the new maritime security regimes for ships, passengers, and cargo, would of necessity be global. "When overnight you go from one per cent of your budget to fifty-eight per cent of your budget being devoted to maritime security; that is a major shift. And it has caused other people to finally wake up and say, 'We haven't properly resourced this agency…' That has led to the support we've seen for funding our ideas on how to secure our maritime homeland." A global reach agency prior to 9/11, Pluta thought that the Coast Guard would now view its daily missions in a permanent global context.

For Captain William "Russ" Webster, who as Chief of Operations in the First District dealt directly with field commanders, the need was immediate for thoughtful policy decisions on the exact nature of the Coast Guard as a maritime homeland security force.

> One of the biggest challenges in the field right now is that the same guy who's out there as a lifesaver, the same guy who today is going on board a recreational boat wearing moccasins so he doesn't scuff the

deck, he may be faced with the dilemma of going to our level six continuum of force—deadly force—on that same vessel, in a very different circumstance, the following day. I don't think it's fair right now to ask our people to have not only the breadth of knowledge they had before about SAR plus some 600 different fisheries regulations, but now a whole new domestic terrorism continuum use of force.

By its adherence to the multi-mission doctrine, the Coast Guard created such conundrums for itself. Webster saw the answer in the creation of new legions trained specifically to act in certain, more military, ways, for maritime homeland security. One way to accomplish this, he speculated, would be to bring the Port Securityman rating, or some combination of security, law enforcement, and military ratings, back into the active duty corps. And the port security dilemma would be played out in an arena where high interest vessels would load and unload cargo on the same waterways where recreational boating was expected to increase by sixty-five per cent over the next twenty years.

Captain Dana Goward, an aviator picked by Admiral Riutta to standardize platforms, training, and doctrine for Coast Guard small boats, saw similar problems. 9/11 triggered a greater emphasis on speed and armament in the Boat Force. Standardizing the Boat Force was job enough, but after 9/11 pressures built to instantaneously arm the small boat stations as well. "We suggested that it would be appropriate to move the automatic weapons from the stations to the Groups or to the district armories, because the folks at the stations were going in a number of different directions, didn't have the time or expertise to maintain the weapons, or to train properly with the weapons. We thought we'd get a decrease in the workload at the stations and an increase in the effectiveness of the automatic weapons, because they'd be maintained by someone who was prepared to maintain them and they'd be with the people who would actually conduct the training when and if it were needed." Commander James D. Maes, an assistant to Captain Goward in the re-engineering of the Boat Force, suggested as did many others that a new generation of weapons were required for small boats, weapons that did not carry the risks of long-range collateral damage in a domestic port environment.

The weapons issue was the tip of a deeper issue. Was maritime anti-terrorism a law enforcement mission or a defense of the homeland mission, and, once that doctrine issue was decided, what level of force was required? The Boat Force, in something of an operational and doctrinal crisis prior to 9/11, took upon itself a staggering number of the Coast Guard's total response hours after the attacks. The increased operational tempo for the small boats only served to magnify these and many other vexing questions. After 9/11, requirements were redrawn for the new 25' Response Boat Small (RBS), which was designed to replace an array of non-standard small boats scattered throughout the Service. 9/11 led to enhancements in the speed of the new boat, as well as to provisions for weapons mounted fore and aft. The same types of design modifications were put in place for the Response Boat Medium (RBM), the anticipated replacement for the workhorse 41' Utility Boat (UTB).

For others, like Vince Patton, the Master Chief Petty Officer of the Coast Guard during 9/11, homeland security was never about the Coast Guard taking on a new mission. Rather, maritime homeland security was already a part of the missions the Coast Guard performed even before 9/11.

"Admiral Loy and many of the Commandants before him have been trying to say that all along, but because prior to 9/11 it was 'out of sight, out of mind' nobody thought we'd be attacked. Go back to when John Cullen found those saboteurs on the beach on Long Island in World War II—clear evidence that if it happened once before, it can happen again. Therefore, we have to be prepared at all times. I dread to think of how we would have survived a big maritime hit soon after 9/11. That question goes through my mind a lot. We would not have been prepared. Thank God it didn't happen."

Captain Mike Lapinski, Admiral Loy's press aide on 9/11, believed that the Coast Guard had everything in place in New York Harbor if 9/11 turned out to be a law enforcement mission, or even if it turned into "extremely low intensity combat. But, if it was something more than that that was planned [by the terrorists] then we certainly would have wanted a [U.S. Navy] Aegis cruiser there." Along those lines, Lapinski recalled the conversations between Admiral Loy and the Chief of Naval Operations, Admiral Verne Clark, as consisting of what the Navy could do for the Coast Guard, and not vice versa.

Lieutenant Chris Kluckhuhn, who commanded one of the Air Station Cape Cod helicopters that sought to rescue people from the twin towers on 9/11, remarked that the rules aviators follow were always in a state of flux, and the new homeland war footing was no exception. "Is an H-60 worth sacrificing to take out a plane that's full of explosives and flying toward a building full of people? Those are the kinds of ethical decisions we're now faced with."

For many, their experiences at Ground Zero and the landfill at Fresh Kills, as well as throughout the nationwide Coast Guard response, only strengthened a sense of the value of their Coast Guard missions. For others, 9/11 was the signal for them to leave the Service, to spend more time watching children grow up. "All I could think of the whole time I was in New York was how much I wanted to hug my kid," remembered a single parent who decided to go on terminal leave six months after 9/11. This Coast Guardsman, who had grown up in a Navy family, was even more shocked that an attack had been made on the Pentagon, the symbol of American power.

After the World Trade Center and the anthrax cases, Coast Guard National Strike Force personnel reflected at length on the meaning of their experiences. They thought how 9/11 might better prepare them for the new and more dangerous battlespace they were entering. "We've done a lot of great things with technology in the Coast Guard," said Lieutenant Commander Nathan Knapp, executive officer of the Atlantic Strike Team on 9/11. "But one of the things that maybe we've sacrificed is how to do it in the absence of technology. And I think that's something we don't want to lose. Here we were in one of the greatest metropolis' the world has ever known, and nobody could make a phone call."

"I don't know that there's a better prepared unit in the Coast Guard," said Knapp's operations officer, Lieutenant Scott Linsky. "We respond to hurricanes, floods, earthquakes, and it's all basically the same: you get someplace and the infrastructure is wiped out. 9/11 was more hectic than anything we'd ever responded to because of the continuously emerging nature of the hazardous materials response. And we brought a bunch of VHF radios to New York, which we should have known wouldn't work amongst all those concrete and steel buildings.

Cell phones didn't work for the first ten days. We had satellite phones, but it would have been nice to have a satellite data link, to enable us to tap into some of the Coast Guard data bases we use, and to enable us to get pictures, e-mails, and sitreps back to Fort Dix. The Pacific Strike Team is prototyping a model like this now. Otherwise, technologically, I think we did okay. After every incident we have a debrief and I ask my staff, what *didn't* we have. If it's something we need to be prepared next time, we get it." Chief Dan Dugery of the Atlantic Strike Team was succinct in what he would like to have next time. "Better air monitoring gear. A portable gas chromatograph mass spectrometer would be nice."

One area the Strike Team looked into after 9/11 was industrial hygiene and safety, searching for someone who could take the air and water monitoring data the team produced and turn it into a comprehensive picture of necessary site safety precautions. Chief Warrant Officer Leonard Rich wished he had twenty minutes of video of the side streets around Ground Zero from the first days after the attacks. It would have been a view of site safety issues unlike any other. With all that, less than a dozen injuries were reported by workers at Ground Zero.

The issue of sending the Strike Teams into the heart of a nuclear, radiological, chemical, or biological attack is also one the team discussed at length. "We're not the fire department," said Lieutenant Linksy. "If anything, we're overly cautious. We always have the opportunity to step back and say, 'I'm not comfortable here, let's reassess and figure out how we're going to do this.'"

Boatswain's Mate First Class David Bittle wanted guns on board the Strike Team's 32' Sea Ark on the night of the 11th. "We were tasked that night and for the next several days with doing security patrols in these rivers, stopping people and turning them around.

And we would stop people and who knew what kind of agenda they had that night and those early days?" On the non-standard 32' boat, which would have to be specially altered to support an M-60, he would have been comfortable with a light weapons suite including 9mm Barettas, 12-gauge Remington 870 shotguns, and M16-A2s. "A lot of the cutters didn't realize we didn't have weapons on board," remembered Bittle. With a name like "Strike Team," the assumption is of a Rambo-like unit. When their lack of armament became clear, the 32-footer was broken off from security work and put to work making logistical runs.

The head of the "M" side of the Coast Guard, Admiral Pluta, addressed the issue this way: "I will always fall down on the side of protecting the health and safety of our people, regardless of what signal of alarm that sends to somebody else. Maybe they should be alarmed. But that shouldn't prevent us from being smart about how we package our response. Calling a decon station a washing station, I have no problem with that. But if it means that, rather than scare somebody, I should go into a hazardous environment without a suit on or without a respirator on, I'm totally against that. We need to deal with the hazards as we know them, and protect our people so they can come back to fight another day."

In terms of the overall budget of the Service, it would be too facile to suggest that Osama Bin Laden saved the U.S. Coast Guard. But there was no denying that after 9/11 the nation suddenly discovered the vulnerability of its ports and waterways, and wanted the Coast Guard to secure them. In the immediate aftermath of 9/11, previous across-the-board cuts were rescinded and the Service's budget increased from five to six billion dollars in less than a year. For Vice Admiral Timothy Josiah, the Chief of Staff before, during, and after 9/11, the effects of the attacks could not have been more dramatic.

> On September 10th, we were about to be forced to decommission two medium endurance cutters way before we wanted to, to decommission some buoy tenders and several C-130 aircraft, helicopters, and Falcon aircraft, *and* take fifteen per cent out of our major deployment cutter days and aircraft hours. It was an across-the-board operational cut,

with no policy behind it with anybody saying that they wanted less Coast Guard out of it. But we couldn't pay our bills and that's what we had to do to get to the answer... The fifteen per cent cut was to be permanent. About two weeks after the attack, we were given two hours to come up with what we needed for a supplemental appropriation. We came up with a number at eight o'clock in the morning. By ten o'clock it had been doubled, from one hundred to two hundred million dollars. By noon, the Department had increased it to almost six hundred million dollars.

Admiral Hull, as he sought closer ties with state agencies and governors in his Ninth District in Cleveland after 9/11, believed that districts themselves would have to be altered, even if slightly, in order to better coordinate the first response capabilities of the Coast Guard with those of states. "For example, Group Detroit takes care of Cleveland. Well, Cleveland doesn't want Group Detroit taking care of it. Now, *Group Lake Erie* could take care of Cleveland. There are some Group/MSO boundaries that need to be looked at. We can make [the pre-9/11 boundaries] work in the Coast Guard, but they don't make sense to somebody we're delivering services to. Every two years, DoD's unified command plan asks if their boundaries match the world they're living in. We've now got to do something [similar], not for budgetary reasons but for delivery of service reasons."

There will be problems in bringing the Activities concept to the whole Coast Guard. The loss of command opportunities with the loss of MSOs and Groups, the opposition of some to the very name 'Activities,' much less the concept, will for a time keep it at bay in some districts. Some O-5s in the Service held command over seven people, while Captain Patrick Harris's Prevention and Compliance Division Chief at Actitivities New York, a non-command position, led 111 people. But the former might be promoted over the latter because of command experience. Yet, for Admiral Bennis, one of the Coast Guard heroes of 9/11, there was everything to be gained from the shift to the Activities concept throughout the Service.

We're the people that always have the day job, in both war and peace. We always have our daily missions, and there are many of them, regardless of whether or not the country is at war. If you take one of those standard 'what the Coast Guard does in a day' posters and apply it to Activities New York, it's a helluva lot. Forty million ferry passengers moving through the harbor every year, the amount of petroleum that moves through, the number of SAR cases. There is no better way to run our operations than the Activities concept. One person owns all the problems, as well as all the assets to solve them.

Admiral Barrett remarked that the situation could also change depending on the mission. 9/11 was a port-centric response. In the future, as Maritime Domain Awareness pushes the Coast Guard farther and farther offshore, the connectivity between the Groups and the MSOs that worked so well on 9/11 might not be appropriate to offshore missions.

Yet there can be no doubt that the model worked in New York. Force lay-down policies were hastily constructed to handle the rapid and massive influx of the Coast Guard cutters, boats, and aircraft that enforced security zones. Traffic management plans were formed and published to keep the maritime community aware of the safety and security zones the Coast Guard had established. And new policies for boarding commercial vessels were developed and circulars describing those policies were disseminated throughout the port.

These three elements defined the operation of Activities New York once the initial evacuation of Manhattan had concluded. Many of these local Activities' policies, formulated in the first hours and days after the attacks, were subsequently validated through adoption across the entire Coast Guard.

Another, perhaps unseen, aspect of an Activities command was in its melding of what often seemed like a hopelessly bewildering array of Coast Guard missions. Besides its regular "day job," missions of port safety and vessel traffic management, Activities New York was able to quickly assimilate Port Security Units for defense-related asset protection, Atlantic Strike Team elements that provided Incident Management System and environmental monitoring skills, and large offshore cutters that brought secure command and control to an initially chaotic situation. The effect on Service cohesion, indoctrination and, by extension, *esprit de corps*, of the joining of these disparate elements into a single complex mission was not to be underestimated.

For Admiral Barrett, the mix of missions had a definite nexus, one that argued against any attempt to break the Coast Guard into different elements and parcel them out to different federal agencies, a stance confirmed in the enabling legislation for the new Department of Homeland Security. "Our ability to put assets on top of a problem anywhere along the U.S. coast, even in Alaska, within an hour or two, that's a huge capability. There isn't another organization in this country, I believe, that has the ability to put people, assets, cutters, aircraft, trained capable resources against a problem as quickly and as well-managed as we do on a nationwide basis."

That ability, argued Admiral Errol Brown and many others, evolved from the Service's core mission to save lives. "Every Coast Guardsman is a lifesaver. That's where it all starts, with the core of our humanitarian service. It's the thing that people come in to do, if not to save lives then to help in some fashion with the environment."

After 9/11, on a basic level, Coast Guard budgets would be bigger, and a bigger percentage of them would be devoted to maritime security. At the strategic level, the new Maritime Security Condition One led to certain inescapable conclusions. For Admiral James Carmichael, Commander of the Seventh District during 9/11, no matter the final structure of any Homeland Security Department or similar construct, that organization would have only one place to go for port security and protection of the waterways of the nation, and that place was the Coast Guard.

How the Service decided to provide security to the ports and waterways—how much would be done by the Coast Guard and how much by other means—was destined to become the core debate within the Service for years to come. And in that debate, the dual role of the Coast Guard in providing both safety and security would continue to defy easy definition.

Acknowledgements

"If you're a military Service you shouldn't have to make excuses for having a band or a history program. It's a part of doing business."

-Rear Admiral Roy J. Casto,
Eighth District Commander on 9/11

The U.S. Coast Guard has not used trained historians to document current operations since 1945. Soon after 9/11, the Assistant Commandant for Government and Public Affairs, Rear Admiral Kevin J. Eldridge, and the Chief of Coast Guard Public Affairs, Captain Jeffrey Karonis, decided that the Coast Guard's response to the terrorist attacks in New York and Washington, as well as the high-level decisions about the future of the Service being made by the senior Service leadership, deserved to be recorded in something approaching real time. In early October, a memo from Secretary of Defense Donald Rumsfeld, asking each of the five Armed Services to document their operations in response to 9/11, added more impetus to the plans of Admiral Eldridge and Captain Karonis.

Soon after, the Captain's Executive Officer, Ms. Pat Miller, asked if I would come on board to document Coast Guard operations and policy decisions on and after 9/11. As a college professor of American Studies and anthropology and a Reserve Chief Petty Officer who has augmented the Coast Guard Historians Office since 1991, I volunteered to come onto active duty for a year under the general Title 10 recall. This year was later extended to eighteen months. By October 16, 2001, the Coast Guard Historian, Dr. Robert Browning, and I had worked up a proposal to document Operation *Noble Eagle* from the Coast Guard point of view. This proposal was approved by Admiral Eldridge later in October and in early December I went to work.

This volume is the sixth and final of the tasks we set out to accomplish under the original proposal approved by our admiral. Looking back at the original proposal, it is interesting now to read that we would collect "files that will allow the study of both immediate and long-term consequences for the organization, from the stand-up of a homeland security office, with its potential to reorient and perhaps restructure USCG operations in both the near and long-term." That bold prediction now seems rather the understatement of the decade.

First, I would like to thank Captain Karonis, who saw the value of this project from every conceivable angle: historical documentation, strategic analysis of lessons learned, Service cohesion and *esprit de corps*, public affairs, recruiting, and the sense many of us had in mid-September that if we didn't get it down on paper, future generations would curse our lack of vision. Pat Miller collapsed more bureaucratic roadblocks than I could imagine being built in a year, and Admiral Eldridge made sure the project was supported with a very strong ALCOAST message on January 14, 2002, and with repeated reminders to his fellow flags to support the effort. "Doc" Browning guided me through my first active duty tour without destroying my sanity or losing his own, a commendable double accomplishment. Others in the Historians Office, Scott Price, Chris Havern, and Auxiliary Staff Officer Jerry Counihan, were always around to sound ideas and share frustrations with.

Having never before been called to active duty, I had a steep learning curve. To name just a few: studying the headquarters boxes and their interrelationships; finding the key elements of operational data necessary for an operational history; learning the operational platforms, their characteristics and performance on 9/11; defining the M/O split within the Service; learning the terminology, and so forth. In this I was helped immensely by dedicated people throughout the organization.

At Coast Guard Headquarters, LCDR Jane Cubbon (G-ORP-3), went out of her way to preserve the work of the HQ 9/11 Incident Management Team (IMT) and make it accessible, along with providing office space for me to work in during my first three weeks at HQ. LCDR Mike Ryan (G-OCC-2) introduced me to CGINFO and made the new daily Abstract of Operations system available. CDR Warren Soloduk and his staff prepared a detailed series of encapsulated timelines of each element of Coast Guard operations after 9/11. CDR Ted Bull of the Commandant's Staff understood what was in front of me better than I did. CDR James McPherson, Chris Havern, Jack O'Dell, and especially Jolie Shifflet, made my deployment to Cuba largely effortless.

Throughout the Service, and in addition to those many personnel mentioned or quoted in the text, a small cadre of operational officers and enlisted in command of or attached to cutters, as well as officers from Activities New York and the Atlantic Strike Team, shared their thoughts and experiences either electronically or in person. Notable amongst these: SCPO Steve Cantrell, USCGC *Ridley*; LT Steve Wittrock, USCGC *Katherine Walker*; LCDR Bill Milne and LT (j.g.) A.E. Florentino, USCGC *Juniper*; LT Ron Catudal, HDC 201; CAPT Lapinski and CDR Bull, COMDT; BMCS Mike Chadwick, USCGC *Pelican*; QM1 Steve Carriere, USCGC *Grand Isle*; LT Brian Fiedler, USCGC *Monomoy*; CDR F.M. Midgette and QM1 Derosier, USCGC *Forward*; CDR Charles Adams, HQ-IMT, and LT Benjamin A. Benson of MSO San Diego. PAC William Epperson at Coast Guard Headquarters preserved over a year's worth of situation reports and other materials related to 9/11. Along with those quoted or cited in this book, these personnel deserve enduring recognition for their care, concern, and high purpose in preserving their part of the record of this seminal event in the history of the U.S. Coast Guard.

LCDR Nathan Knapp, CWO Leonard Rich, and MSTC Dan Dugery of the Atlantic Strike Team went out of their way to make that entire unit available for oral histories, as well as transporting me to Freshkills and Ground Zero. Several retired Coast Guard personnel intervened on behalf on the project at several points on the route, and to them—Vice Admiral Howard Thorsen, MSTCS Dennis Noble, LCDR Tom Beard—I offer my profound thanks.

The senior leadership of the Service responded on the whole

extremely well to having a Reserve Chief nosing into their offices and files. During the course of this project, I was able to visit each Coast Guard district for oral history interviews of each District Commander, both Area Commanders, and the senior leadership of the Service (Commandant, Vice-Commandant, Chief of Staff, and the Assistant Commandants in charge of Deepwater, Operations, Marine Safety, and Reserve, as well as the MCPO-CG) at Coast Guard Headquarters, as well as the commander of the U.S. Navy Command Center at the Pentagon. This operational documentation involved twenty-eight flights, over 30,000 miles of air travel, and more than 2,000 miles of ground travel. In addition, I gathered oral histories at several units, including the Atlantic Strike Team at Ft. Dix, NJ; Activities New York based at Ft. Wadsworth, NY; USCGC *Bainbridge Island*; USCGC *Adak*; USCGC *Hawser*; USCGC *Bear*; Harbor Defense Command Unit 201 in Newport, RI; and Air Station Cape Cod. These operational documentation missions resulted in over fifty hours of oral histories on digital video tape. These were subsequently transferred to permanent storage medium on DVD and transcribed for permanent historical research use. Situation reports, unit paperwork, and abstracts of operations were also collected, resulting in an archive of 9/11 operations consisting of several thousand pages of primary historical data. We also had the advantage of half a dozen oral history interviews of Coast Guard personnel conducted in New York in early October by the 126th Military History Detachment of the Massachusetts Army National Guard.

Even given this mass of material, however, a complete picture of the complexity of the U.S. Coast Guard's organizational response to 9/11 is far beyond the scope of the Coast Guard's historical collection and analysis apparatus as currently constituted. On the positive side, this very challenge triggered a long-overdue evaluation of the institutional weakness of USCG historical data collection and writing. Captain Karonis was forthright in his acceptance of the need for such an evaluation, and actively sought opinions and potential solutions from within the organization, from retirees, and most notably from firsthand visits to the history programs of the other uniformed Services. In these latter visits, I would like to thank especially Brigadier General David Armstrong, USA (Ret.), Dr. Mickey Schubert, and Dr. Hans Pawlisch of the Office of the Chairman of the Joint Chiefs of Staff; Dr. William Dudley, Director of the Naval Historical Center; Dr. Richard Hallion and Chief Master Sergeant Walt Grudzinskas of the Office of the Air Force Historian; Dr. Charles D. Melson of the History and Museums Division of the U.S. Marine Corps; Dr. William Epley of the Center for Army History; and Major Robert Smith, Commander of the 305th Military History Detachment at Fort Lesley J. McNair. I owe many thanks to Captain Ray Pietrzak, USNR, for his leadership of our Joint History Team expedition to document Coast Guard and Navy participation in JTF-160 port security operations at Guantanamo Bay, Cuba.

Finally, I thank my wife and son and daughter for their exemplary patience during my long absences for both deployments and writing. As always, they are my own personal homeland security.

Despite all of these assists, I assume responsibility for all errors of fact or interpretation in this document. I apologize in advance, as well, that this one-year project was only able to produce what those who were there will realize is but a snapshot of Coast Guard operations and policy decisions on and after 9/11. The full picture may never be drawn, but we are in a far better position to attempt that full portrait than we were before this ambitious project was approved and supported by Admiral Eldridge.

-P.J. Capelotti, PAC, USCGR, Ph.D.
Washington, DC
January, 2003

Appendix

Oral History Interviews*

<u>Flag corps/senior leadership</u>
ADM James M. Loy, Commandant, USCG HQ
VADM Thomas H. Collins, Vice-Commandant, USCG HQ
VADM Timothy W. Josiah, Chief of Staff, USCG HQ
VADM Ernest "Ray" Riutta, PACAREA, Alameda, CA (interviewed in Juneau, AK)
VADM Thad W. Allen, LANTAREA, Portsmouth, VA
RADM George N. Naccara, Commander, D1, Boston
RADM Richard Bennis,USCG (Ret.) (interviewed in Washington, DC)
RADM Jeffrey Hathaway, CO, U.S. Navy Command Center, Pentagon
RADM Patrick M. Stillman, G-D, USCG HQ
RADM Paul J. Pluta, G-M, USCG HQ
RADM Terry M. Cross, G-O, USCG HQ
RADM Carlton D. Moore, G-WT, USCG HQ
RADM Dennis Sirois, G-WT, USCG HQ
RADM James S. Carmichael, Commander, D7, Miami
RADM Roy J. Casto, Commander, D8, New Orleans
RADM James D. Hull, Commander, D9, Cleveland
RADM Errol M. Brown, Commander, D13, Seattle
RADM Ralph D. Utley, Commander, D14, Honolulu
RADM Thomas J. Barrett, Commander, D17, Juneau (interviewed in Washington, DC)
MCPO-CG Vince Patton, USCG HQ

<u>Other District Officers</u>
CAPT William "Russ" Webster, D1 Operations
CAPT Paul D. Kirkpatrick, D7 Maritime Homeland Security
CAPT James W. Stark, D7 Operations
CAPT Joel R. Whitehead, Chief of Staff, D8
CAPT Richard Sullivan, D8 Maritime Homeland Security
CAPT Thomas D. Yearout, Chief of Staff, D14
CAPT Steven A. Newell, D14 Operations
CAPT Terry L. Rice, D14 Marine Safety
CDR Michael J. Scully, D7

<u>Atlantic Strike Team</u>
CDR Gail P. Kulisch, CO, Atlantic Strike Team (AST)
LCDR Nathan E. Knapp, XO, AST
LT Scott R. Linsky, OPS, AST
LT (j.g.) David Reinhard, AOPS, AST
LT (j.g.) Christopher Williammee, AST
CWO Leo Deon, Boatswain, AST
CWO Paul Johnston, AST
CWO Leonard Rich, Engineer, AST
MSTCS Dean E. Matthews, AST
MSTC Dan Dugery, AST
MST1 John Kapsimalis, AST
MST 1 Robert J. Schrader, AST
MST2 Monica L. Allison, AST
BMC Robert W. Field, Jr., AST
BM1 David M. Bittle, AST
BM1 Patrick G. McNeilly, AST
YN1 Matthew T. Leahy, AST
YN1 Grant Smith, AST
MK1 Charles G. Nowak, AST
MK2 Robert Cummins, AST

<u>Auxiliary</u>
Staff Officer Ellen Voorhees, Division Seven, USCG Auxiliary
Staff Officer Tom Murray, Fifth North Division, Flotilla 7-11, USCG

Auxiliary

<u>Public Affairs</u>
CAPT Jeffrey Karonis, G-IPA, USCG HQ
CAPT Michael J. Lapinski, PAO, G-C, USCG HQ
LCDR Brendan C. McPherson, PAO, LANTAREA, Portsmouth, VA

<u>Air Station Cape Cod</u>
CAPT Richard P. Yatto, CO, Air Station Cape Cod (AirStaCC)
LT Joseph R. Palfy, Royal Canadian Air Force, AirStaCC
LT Christopher L. Kluckhuhn, AirStaCC
LT Kurt R. Kupersmith, AirStaCC

<u>Activities New York</u>
CAPT Patrick A. Harris, Deputy CO, Activities New York (ACTNY)
CDR Michael F. McAllister, ACTNY
LCDR Kevin J. Gately, ACTNY
LT Michael H. Day, ACTNY
BM1 Kenneth Walberg, ACTNY
BM3 Carlos Perez, ACTNY

<u>Cutter Force</u>
CAPT Dan Deputy, USCG HQ, G-OCU
CDR Frank M. Reed, USCGC *Bear* (interviewed in Guantanamo Bay, Cuba)
LT Sean C. MacKenzie, CO, USCGC *Adak* (interviewed at Station Sandy Hook, NJ)
LT Christopher Randolph, CO, USCGC *Bainbridge Island* (interviewed at Station Sandy Hook, NJ)
BMC James A. Todd, POIC, USCGC *Hawser* (interviewed ay ACTNY)

<u>Boat Force</u>
CAPT Dana Goward, USCG HQ, G-OCS
CDR James D. Maes, USCG HQ, G-OCS-2

<u>Harbor Defense Command Unit 201</u> (interviewed in Newport, Rhode Island)
CAPT Joanne F. Spangenberg, CO, Harbor Defense Command Unit 201 (HDC 201)
LT Ron J. Catudal, HDC 201
SCPO Bradley Blatchley, HDC 201

<u>Port Security Unit 305</u> (interviewed at Guantanamo Bay, Cuba)
CDR R.W. Grabb, CO, Port Security Unit 305 (PSU 305)
LCDR Lee A. Handford, XO, PSU 305
LCDR Karl S. Leonard, Ops, PSU 305
DC3 Eric Bowers, PSU 305
BM3 Michael Walker, PSU 305

<u>Civilians</u>
Dr. Alan S. Schneider, USCG HQ
Ms. Frances Townsend, G-C2, USCG HQ

Interviews conducted by the 126th Military History Detachment, Massachusetts National Guard (conducted in New York, October 2-6, 2001)
CDR Gary M. Smialek, USCGC *Tahoma*
LT Leona Roszkowski, PSU 305
ENS Erika J. Lindberg, USCGC *Tahoma*
BMC Walter Haven, PSU 305
BM1 Mark Baumgaetner, National Strike Force
YN2 Chad Pollack, National Strike Force

*This list of Coast Guard personnel who offered oral history interviews for the Operation Noble Eagle Documentation Project shows rank and billet at the time of the interview.

Notes

Unless otherwise noted, all oral history/operational documentation interviews were conducted by CPO P.J. Capelotti as part of the Operation *Noble Eagle* Documentation Project. This effort began on December 15, 2001, with the collection of situation reports and other primary materials related to 9/11. Interviews with selected personnel began in February, 2002, with a four day visit to the Atlantic Strike Team at Fort Dix, New Jersey.

During the next six months, more than eighty oral history interviews were collected, during a period of time when the entire structure and future of the U.S. Coast Guard was in a state of flux. Many senior flag officer interviews were conducted prior to the proposal by the Bush Administration to create a Department of Homeland Security. Others, like that with Vice Admiral Riutta, were conducted after this announcement (and in the case of Admiral Riutta the interview could not be scheduled until August, 2002, two months after his retirement from the Service). The schedule of interviews, undertaken as the Service tried to find its post-9/11 identity, is delineated in the following Notes, as it effected the questions asked and perhaps the answers offered.

This history was written during the fall of 2002 and submitted to the editorial advisory board of the Coast Guard Historians Office on December 8, 2002. The manuscript was approved for publication in mid-January, 2003, by Rear Admiral Kevin J. Eldridge, Assistant Commandant for Government and Public Affairs (G-I). Design and lay-out was done by Chief Warrant Officer Lionel M. Bryant in late January, 2003. Final corrected transcriptions of the oral histories are still in progress, as is the archiving of the primary data related to

9/11. These will all be incorporated in the 9/11 Archive of the Coast Guard Historians Office at Coast Guard Headquarters.

Prefaces/Pentagon and 41497

These two prefaces are based on oral history interviews conducted at the Pentagon with Rear Admiral Jeffrey Hathaway on June 20, 2002 and at Activities New York with Boatswain's Mate Third Class Carlos Perez on April 4, 2002. The statistic regarding the use of the Coast Guard's small boats in the weeks after 9/11 was cited to the author by several officers, including First District commander Rear Admiral G.N. Naccara. "We'll be paying for that for many years," the Admiral noted. "But we could not quantify the amount of time and wear on our people as well as we could on our boats and aircraft, so that could really have a long-term effect on people."

One: 9/11 at Activities New York:
Richard Bennis and the Priorities of Port Security

This chapter is based on an oral history interview conducted with Rear Admiral Richard Bennis, USCG (Ret.), in his office at the new Transportation Security Administration on May 31, 2002. Additional materials came from interviews at Activities New York on April 4, 2002 with CAPT Patrick A. Harris, CDR Michael F. McAllister, LT Michael H. Day, BM1 Kenneth Walberg, BM3 Carlos Perez, and BMC James A. Todd, POIC, USCGC *Hawser*. LT Sean C. MacKenzie, CO, USCGC *Adak*, and LT Christopher Randolph, CO, USCGC *Bainbridge Island*, were interviewed at Station Sandy Hook, NJ, on May 6, 2002, as was LCDR Kevin J. Gately, USCGR, who supported the collection, writing, and publication of Activities New York's After-Action Report entitled *"Guarding Liberty,"* published under the signature of Admiral Bennis on 15 March 2002. This monumental primary document includes a standard after action report, as well as chapters on best practices/lessons learned; supervisors after action reports; debriefing reports; and feedback summary reports. Among other conclusions, Admiral Bennis in his executive summary noted that the command had "warned in the OPSAIL 2000 After-Action Report that the closure of Governors Island and steep downsizing of Coast Guard forces in NY during the past decade seriously compromised the Coast Guard's surge capability in New York, and this has been borne out. While the Coast Guard found the massive amounts of emergency funding it needed, money could not remedy a lack of small boats overnight, or bring forth qualified boat crews and boarding officers who simply did not exist. If anti-terror-ism operations are to be part of the "new normal[cy]," a major increase in funding, resources, and personnel is needed." (Bennis to Commander, First Coast Guard District, March 15, 2002). A copy of *Guarding Liberty* is on file in the 9/11 Archive of the Coast Guard Historians Office at Coast Guard Headquarters.

The number of people evacuated by water from lower Manhattan during 9/11 has been variously estimated from several hundred thousand to over a million, with the majority of estimates somewhere in between. The combination of mass confusion, people fleeing north along the streets of New York, and the eventual reopening of New York's bridges to a pedestrian evacuation, renders an accurate picture of the waterborne evacuation problematic. The figure most often cited by Coast Guard personnel who were there that day is around 750,000.

Two: 9/11 and the Atlantic Strike Team:
Marine Safety at Ground Zero and Fresh Kills

This chapter is based on a series of oral history interviews conducted with members of the Coast Guard's Atlantic Strike Team at Fort Dix, New Jersey, from February 19-22, 2002. LCDR Nathan Knapp provided access to unit members and a quiet place where the interviews could be conducted. CWO Leonard Rich went out of his way to show the author both Ground Zero and Freshkills. MSTC Dan Dugery provided a detailed tour of the unit and its gear. Others who provided interviews were CDR Gail P. Kulisch, CO, LT Scott R. Linsky, LT (j.g.) David Reinhard, LT (j.g.) Christopher Williammee, CWO Leo Deon, Boatswain, CWO Paul Johnston, MSTCS Dean E. Matthews, MSTC Dan Dugery, MST1 John Kapsimalis, MST 1 Robert J. Schrader, MST2 Monica L. Allison, BMC Robert W. Field, Jr., BM1 David M. Bittle, BM1 Patrick G. McNeilly, YN1 Matthew T. Leahy, YN1 Grant Smith, MK1 Charles G. Nowak, MK2 Robert Cummins, Staff Officer Ellen Voorhees, Division Seven, USCG Auxiliary, and Staff Officer Tom Murray, Fifth North Division, Flotilla 7-11, USCG Auxiliary.

Copies of CWO Rich's site safety plans for New York (Ground Zero) and Fresh Kills Landfill are on file in the 9/11 Archive of the Coast Guard Historians Office at Coast Guard Headquarters, along with a small number of additional primary materials from the National Strike Force response to 9/11.

Three: 9/11 at First Coast Guard District in Boston:
George Naccara and the Northeastern Maritime Frontier

This chapter is based on an oral history interview conducted with Rear Admiral George N. Naccara, USCG (Ret.), in his study at First Coast Guard District Headquarters in Boston on April 2, 2002. Captain William R. Webster, D1 Operations, was also interviewed at this time. Earlier that same day, interviews were conducted at Air Station Cape Cod with CAPT Richard P. Yatto, CO; as well as LT Joseph R. Palfy, Royal Canadian Air Force; LT Christopher L. Kluckhuhn, and LT Kurt R. Kupersmith. Prior to the interview with Admiral Naccara, Dr. Alan S. Schneider, a civilian employee of the Coast Guard in Washington, DC, provided technical background material on the nature of Liquefied Natural Gas and its transportation on board LNG tanker ships.

Four: 9/11 and the Area Commanders:
Thad Allen (LANTAREA) as Resource Broker;

Ray Riutta (PACAREA) and the Centrality of Intelligence to <u>Warfighting</u>

The oral history interview with Vice Admiral Thad William Allen took place in his LANTAREA Headquarters office in Portsmouth, Virginia, on March 22, 2002. In addition to this interview, the Admiral provided copies of documents and correspondence surrounding the many tactical issues he faced, as well as copies of his public messages and a copy of his comprehensive historical work, *The Evolution of Federal Drug Enforcement and the United States Coast Guard's Interdiction Mission* (Massachusetts Institute of Technology, 1988). This material is on file in the 9/11 Archive of the Coast Guard Historians Office at Coast Guard Headquarters. LCDR Brendan C. McPherson, PAO, LANTAREA, also provided valuable insight into the Area's handling of 9/11.

The oral history interview with Vice Admiral Ernest "Ray" Riutta was conducted at his office at the Alaska Seafood Marketing Institute in Juneau, Alaska, on 20 August 2002, about two months after the Admiral's retirement from the Coast Guard. Background information on his LORAN station work in Vietnam can be found in *The Coast Guard at War*, by Alex Larzelere (Naval Institute Press, 1997), pp. 205-208. A chapter on the Lampang, Thailand, station, and Thad Allen's time there, can be found in the same work on pp. 275-279.

Five: 9/11 and the Office of the Commandant: <u>James Loy as Strategic Field Commander</u>

Much of this chapter is based on an oral history interview conducted at the Office of the Commandant with Admiral James Loy on 27 March 2002. Further interviews related to Headquarters' handling of 9/11 were provided by MCPO-CG Vince Patton (March 21, 2002), VADM Thomas H. Collins (April 24, 2002), CAPT Jeffrey Karonis (March 26, 2002), and CAPT Michael J. Lapinski (March 26, 2002). The response of the Office of the Chief of Staff was detailed through an interview with VADM Timothy W. Josiah on April 8, 2002. Additional detail provided by Captain Glenn A. Wiltshire (personal communication), Executive Assistant to the Assistant Commandant for Marine Safety and Environmental Protection, who with Mr. Jeff High (G-MW) attended most of the meetings with the Commandant and other senior Coast Guard personnel. Invaluable insights into senior leadership decision-making and staff operations were offered through personal communications and editorial comments by retired Vice Admiral Howard Thorsen.

For the comments of Loy and Patton on the impact on the service of the *Storis* event, see The Coast Guard Fiscal Year 2002 Budget Request (107–16), Hearing Before The Subcommittee On Coast Guard And Maritime Transportation Of The Committee On Transportation And Infrastructure, House Of Representatives, One Hundred Seventh Congress, First Session, May 3, 2001, archived on the web at:

http://commdocs.house.gov/committees/Trans/hpw107-16.000/hpw107-16_0.htm

The U.S. Commission on National Security/21st Century (the Hart-Rudman Commission), warned in April, 2001, in its final (Phase Three) report, that "A direct attack against American citizens on American soil is likely over the next quarter century. The risk is not only death and destruction but also a demoralization that could undermine US global leadership." Hart-Rudman directly affected the Coast Guard when it recommended combining federal agencies like the Coast Guard, Border Patrol, Immigration and Naturalization Service, and Customs, into a kind of super border patrol service. For the complete texts of the commission's work, see:

http://www.nssg.gov/News/Hart-Rudman/hart-rudman.htm

The full text of the article written for *Homeland Defense Journal* by Admiral Loy and Captain Ross can be found at:

http://www.e11th-hour.org/security/homeland.strategy.html

The new notice of arrival procedures that placed additional restrictions on vessel movements toward, in, and away from U.S., ports, were detailed in Marine Safety Information Bulletin 07-01, dated September 23, 2001. These new restrictions were published in the October 4, 2001, issue of the Federal Register, and were further amplified in Marine Safety Information Bulletin No. 09-01 published on October 5, 2001.

Six: 9/11 and the Coast Guard at the Pentagon: Jeffrey Hathaway and the Navy Command Center

Much of this chapter is based on an oral history interview conducted at the Pentagon with Rear Admiral Jeffrey Hathaway on June 20, 2002. By the time this interview was given, the Navy Command Center at the Pentagon was once again operational, relocated much deeper into the building's maze, while the N3/N5 spaces were still under renovation. Hathaway himself had traveled to ports and Navy units and facilities around the world, studying force protection issues and offering recommendations for what is after all a massive global naval expeditionary force. At one point he found himself amongst a Coast Guard port security unit stationed in a Middle Eastern country, a happy meeting which led to a happier dinner amongst them all.

Even though a Coast Guard admiral, during his Pentagon assignment Hathaway worked for the Navy. Yet within the Navy he sensed some unease that a Coast Guard admiral was in charge of Navy force protection. His uniform led some to assume that the Coast Guard provided force protection for the Navy, which it does not. Even so, Hathaway saw the post-9/11 relationship between the Navy and the Coast Guard, growing closer. How this played out, against the backdrop of such new entities as the defense-focused Northern Command and the homeland security-focused Homeland Security Department remained to be seen. What seemed clear was that the Navy, an expeditionary force fighting a global war on terrorism, had little time nor resources for port security issues within the American homeland. Such issues will remain the province of the Coast Guard. And Admiral Hathaway will return to the Coast Guard from his singular tenure at the Pentagon with an unprecedented understanding of both organizations as they navigate post-9/11 waters.

Seven: 9/11 and the Headquarters Directorates of Marine Safety and Environmental Protection, and Operations:
Paul Pluta, Terry Cross and the "M" and "O" dichotomy

The oral history interview with Rear Admiral Paul J. Pluta was conducted in his conference room at Coast Guard Headquarters on May 3, 2002. The interview with Rear Admiral Terry M. Cross took place in his office on April 9, 2002. Amplification of the issues surrounding the Coast Guard's new responsibilities in maritime intelligence was provided by Ms. Frances Townsend (G-C2), Assistant Commandant for Intelligence, on January 22, 2003.

Eight: 9/11 and the Atlantic Area District Commanders:
Carmichael (D7), Casto (D8), and Hull (D9)

This chapter is based on oral history interviews conducted in the offices of each LANTAREA district commander (except for VADM Allen, the Fifth District Commander who was interviewed in his role as LANTAREA Commander, and RADM Nacarra, interviewed for a separate chapter on the First District, which took the brunt of the attack on 9/11). Rear Admiral James S. Carmichael, Commander, D7, was interviewed in Miami on May 1, 2002. Providing further perspective on the 9/11 response in the Seventh were CAPT Paul D. Kirkpatrick, USCGR, D7 Maritime Homeland Security; CAPT James W. Stark, D7 Operations; and CDR Michael J. Scully, all of whom sat in on the interview with Admiral Carmichael. Rear Admiral Roy J. Casto, Commander, D8, was interviewed in New Orleans on April 11, 2002, along with his Chief of Staff, CAPT Joel R. Whitehead, and CAPT Richard Sullivan, USCGR, Chief of Maritime Homeland Security for D8. Rear Admiral James D. Hull, Commander, D9, was interviewed in Cleveland on

April 26, 2002.

Nine: 9/11 and the Pacific Area District Commanders: <u>Brown (D13), Utley (D14), and Barrett (D17)</u>

This chapter is based on oral history interviews conducted in the offices of each PACAREA district commander (except for VADM Riutta, the Eleventh District Commander who was interviewed separately, in his role as PACAREA Commander). Rear Admiral Errol M. Brown, Commander, D13, was interviewed in Seattle on May 14, 2002. Rear Admiral Ralph D. Utley, Commander, D14, was interviewed in Honolulu on May 16, 2002, along with his Chief of Staff, CAPT Thomas D. Yearout; CAPT Steven A. Newell, D14 Operations, and CAPT Terry L. Rice, D14 Marine Safety. Rear Admiral Thomas J. Barrett, Commander, D17, was interviewed in his office in Washington, DC, on July 2, 2002, after his promotion to Vice Admiral and selection as Vice Commandant.

Ten: 9/11 and the Reserve and Auxiliary: <u>PSU 305 in New York Harbor and Guantanamo Bay, Cuba</u>

This chapter is based on oral history interviews conducted in Newport, Rhode Island, on April 3, 2002, with members of Harbor Defense Command Unit 201. These included CAPT Joanne F. Spangenberg, CO; LT Ron J. Catudal, and SCPO Bradley Blatchley. Interviews with members of Port Security Unit 305 were conducted at Guantanamo Bay, Cuba, during the week of May 20-25, 2002. These included CDR R.W. Grabb, CO; LCDR Lee A. Handford, LCDR Karl S. Leonard, DC3 Eric Bowers, and BM3 Michael Walker. LCDR Leonard, the Unit's Operations Officer, was especially helpful in providing logistics arrangements, transportation, and background information on the Coast Guard's role in detainee operations. Additional perspectives on the role of the Reserve during and after 9/11 were provided by Rear Admiral Carlton D. Moore, Rear Admiral

Dennis Sirois, and LCDR Jane Cubbon. Admiral's Loy's ALCOAST 502/01 asking for a Reserve draw-down was issued on November 1, 2001. Data on Reserve strength prior to 9/11 was provided by LCDR Darrell Prather (personal communication, January 9, 2002). Discussion of the stabilization of the Reserve call-up is contained in "Reserve Mobilization correspondence" located in the 9/11 Archive at the Historians Office.

Data related to the Auxiliary participation in Operations *Noble Eagle* and *Patriot Readiness* was provided by Auxiliary Historian C. Kay Larsen, as well as the Office of the Chief Director of Auxiliary (G-OCX). This data is contained in the "Auxiliary data" in the 9/11 Archive. Further information was derived from "Ground Zero: New York City Auxiliarists and the World Trade Center attack," by C. Kay Larson, *The Navigator*, Vol. 29, No. 3, Fall 2001.

Conclusions: The Coast Guard and the New War

In addition to comments from individuals cited in previous chapters, these conclusions are based on two oral history interviews conducted with Rear Admiral Patrick M. Stillman at Coast Guard Headquarters, the first on March 4, 2002, and the second, soon after the award of the Deepwater contract, on June 27, 2002. Perspectives on the Cutter Force were provided by CAPT Dan Deputy (G-OCU) at Coast Guard Headquarters on April 18, 2002, and by CDR Frank M. Reed, Commanding Officer of USCGC *Bear* (WMEC 901) in Guantanamo Bay, Cuba, on May 22, 2002. Perspectives on the Boat Force were offered by CAPT Dana Goward (G-OCS) and CDR James D. Maes (G-OCS-2) at Coast Guard Headquarters on April 17, 2002.

Index

ISBN 0-16-069824-3